SCHRÖDINGER'S
KITTENS

SCHRÖDINGER'S KITTENS AND THE SEARCH FOR REALITY

SOLVING THE QUANTUM MYSTERIES

JOHN GRIBBIN

LITTLE, BROWN AND COMPANY
BOSTON NEW YORK TORONTO LONDON

First American Edition

ISBN 0-316-32838-3

Library of Congress Catalog Card Number 95-75652

10 9 8 7 6 5 4 3 2 1

MV–NY

Published simultaneously in Canada by Little, Brown & Company
(Canada) Limited

Printed in the United States of America

CONTENTS

ACKNOWLEDGEMENTS

Writing a book like this depends upon the good will of a large number of scientists who have provided me with copies of their scientific papers, often in advance of publication. All these sources of information are mentioned in the text, but some of my correspondents should be given a special mention for the way discussions and correspondence with them have influenced the development of my ideas about quantum reality. In alphabetical order, I would like to thank especially: Bruno Augenstein, of RAND in Santa Monica; Shu-Yuan Chu, of the University of California, Riverside; John Cramer, of the University of Washington, Seattle; Paul Davies, of the University of Adelaide; Dipankar Home, of the Bose Institute in Calcutta; Geoff Jones, of the University of Sussex; Martin Krieger, of the University of Southern California, Los Angeles; and Thanu Padmanabhan, of the Tata Institute in Bombay.

The University of Sussex provided even more help than with my other books by making me a Visiting Fellow in Astronomy and providing me with improved access to an excellent scientific library and to the Internet, while the astronomers at Sussex acted as sounding-boards for some of my less conventional ideas. Without all of these people, this book would not exist.

PREFACE

When I wrote my historical account of the development of quantum theory, published just ten years ago, I never imagined that I would return to the theme of quantum mysteries in another book. In writing *In Search of Schrödinger's Cat* I set out to show just how strange and mysterious the subatomic world of quantum physics is, and the impeccable logic, driven by bizarre experimental results leading to non-commonsensical theories that were in turn confirmed by further experiments, that had forced physicists to take such bizarre notions seriously. The bottom line, as of the mid-1980s, was that for all its strangeness quantum theory works – it is the theory that underpins our understanding of the behaviour of lasers, computer chips, the DNA molecule, and much more besides. The old ideas, so-called 'classical' physics, simply cannot explain such phenomena. What mattered, I stressed in *In Search of Schrödinger's Cat*, was not that the quantum theory was hard to understand, but that, indeed, it worked. The fact that, in the words of Richard Feynman, '*nobody* understands quantum theory' meant that I could conclude my earlier book with the unashamed statement that 'I am happy to leave you with loose ends, tantalizing hints, and the prospect of more stories yet to be told.'

But while I was happy to leave the loose ends dangling, many physicists were not content to rest on their laurels. Not happy with a theory that could not be understood, even if it did work, they have tried strenuously, since I last took stock of the situation in 1984, to solve the quantum mysteries. Along the way, they have made some of the mysteries look even more mysterious and have uncovered new aspects of the strangeness of the quantum world. They have developed explanations of the quantum mysteries that seem to an outside observer to be increasingly bizarre counsels of despair. But they have also, within the past few years, come up with an explanation of the quantum mysteries that just might, after more than sixty years of trying, provide a genuine insight into what is going on – an understanding intelligible not just to the cognoscenti but to anyone interested in the nature of reality.

This new understanding rests not only upon the appropriate interpretation of quantum theory, but also upon the explanation of the behaviour of light within the framework of Albert Einstein's theory of relativity. In this book, I bring both stories up to date, and show that the best explanation of the way the Universe works, the resolution of all the quantum mysteries, requires bringing together quantum ideas and those of relativity theory.

You will not find much of the historical background of the development of quantum theory here; I have covered that ground already. I start with quantum theory as an established success, and discuss some new puzzles and some new ways of looking at old puzzles, before explaining how those puzzles can be solved. You *will* find here everything you need to know to understand what the quantum debate is all about, whether or not you have read anything on the subject (let alone my own books) already; you will read about such seemingly paradoxical phenomena as photons (particles of light) that can be in two places at the same time, atoms that go two ways at once, how time stands still for a particle moving at light speed, and a serious suggestion that quantum theory may offer a way to achieve Star Trek-style teleportation.

To set the scene, though, I begin more or less where *In Search of Schrödinger's Cat* left off, with the famous cat herself, and John Bell's proof that if quantum entities are once part of a single system they remain linked, somehow aware of each other, even when they are far apart. Einstein called this 'spooky action at a distance'; it is more respectably described as 'non-locality'. The concepts may be new to you, or you may think that they are familiar. The 'paradox' of Schrödinger's cat, alive and dead at the same time, has become almost a cliché over the past ten years. But wait. Even if you think you know what it is all about, be prepared to think again. You ain't seen *nothing* yet. I have bigger and better paradoxes, backed up by impeccable experimental tests, with which to amaze you. But they all boil down to one thing. How *can* an electron, for example, go both ways at once through an experiment with two holes? How does it 'know' the structure of the entire experiment at one moment in time?

The total strangeness of the quantum world, the problem we have to solve, can be most clearly understood by looking at the adventures of the twin offspring of our original cat, the kittens of my title. We then have to reconsider just what it is we know about the nature of light itself, the phenomenon that is a key ingredient of both quantum theory and relativity theory. And *then* I will be in a position to point you towards the new ideas that explain the nature of reality and solve the quantum mysteries – *all* of the quantum mysteries. For the first time since quantum theory emerged in the middle of the 1920s, it is possible to say with some confidence what quantum theory *means*. And if that isn't a good enough reason for writing this book, I don't know what is!

John Gribbin
April 1994

All these fifty years of conscious brooding have brought me no nearer to the answer to the question 'What are light quanta?' Nowadays every Tom, Dick and Harry thinks he knows it, but he is mistaken.

ALBERT EINSTEIN
Letter to M. Besso, 1951

No physical world exists behind the apparent elementary sense impressions subjected to the reflection of the mind.

GEORGE BERKELEY
Treatise Concerning the Principles of Human Knowledge, 1710

There are nine and sixty ways of constructing tribal lays, And every single one of them is right!

RUDYARD KIPLING
In the Neolithic Age, 1895

PROLOGUE

The Problem

The central mystery of quantum theory is encapsulated in the experiment with two holes. Don't take my word for it; Richard Feynman, the greatest physicist of his generation, said so on the first page of Chapter One of the volume of his famous *Lectures on Physics* devoted to quantum mechanics.[1] Contrasting quantum physics with the classical ideas of Isaac Newton and the scientists who followed in his footsteps, he said that this phenomenon 'is impossible, *absolutely* impossible to explain in any classical way'. It 'has in it the heart of quantum mechanics. In reality, it contains the *only* mystery.' In another book, *The Character of Physical Law*, he wrote: 'Any other situation in quantum mechanics, it turns out, can always be explained by saying, "You remember the case of the experiment with the two holes? It's the same thing." ' So, like Feynman, I begin with the experiment with two holes, laying bare at the start the central mystery in all its glory. The experiment may seem familiar, but this is one case where familiarity can never breed contempt. The more you know about the experiment with two holes, the more mysterious it seems.

If you encountered the experiment in the school physics lab, it probably didn't seem mysterious at all. That is because nobody bothered (or dared) to explain its mysteries to you; instead, almost certainly, all you were taught was that the behaviour of light passing through two narrow slits in a board, and forming a pattern of light and dark stripes on a screen, was simply a rather neat proof of the fact that light moves like a wave.

As far as it goes, this is true. But it is by no means the whole truth.

---

[1] For my purposes, the terms 'quantum theory', 'quantum physics' and 'quantum mechanics' are interchangeable; full references to books mentioned in the text are given in the Bibliography.

THE LIGHT FANTASTIC

The classic example of a wave is what you see on the surface of a still pond when you drop a single pebble into the water. The waves form a series of ripples, moving outward in a circle from the point where the pebble dropped. If waves like this arrive at a barrier which has just two holes in it, each much smaller than the wavelength of the ripples, then the waves spread out on the other side of the barrier in two semicircles centred on the two holes. The pattern they make is like one half of the pattern of ripples you would get if you dropped two pebbles into the still pond at the same time.

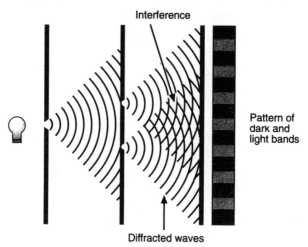

Interference

Pattern of dark and light bands

Diffracted waves

Figure 1
The uniform light from the first hole makes waves which move out in step from each of the holes in the second screen. They interfere to produce a distinctive pattern of light and shade on the viewing screen – proof positive that light travels as a wave.

Everyone knows what kind of a pattern that is. Drop two pebbles into the pond, and you wouldn't actually see two circular sets of ripples passing through each other, but a more complicated pattern of ripples, caused by the interference of the two circular patterns with one another. In some places, the two sets of ripples add up to make extra-large ripples; in other places, the two sets of ripples cancel out, leaving little or no wave motion in the water.

When light is shone through two holes in a board, to make a pattern on a screen on the other side of the board, exactly the same thing happens. For the effect to show up most clearly, it is best to use just one pure colour of light, which corresponds to a single wavelength. The two sets of light waves spread out from the two holes like ripples on a pond, and when the light arrives at the screen it shows a pattern of light and dark stripes (interference fringes) corresponding to places where the waves add together (constructive interference) and places where the waves cancel each other out (destructive interference). All simple, straightforward schoolday science, from which it is quite easy to work out the wavelength of the light, not merely the fact that light is a wave, by measuring the spacing of the interference fringes.

But even at this level, there are subtleties involved. The pattern you get on the screen is *not* the pattern you would get if you let light through each of the two holes in turn and added up the brightness of the two patches of light thrown on the screen. This is one of the key features of how interference works. With just one hole open, you would get a patch of light on the screen just behind that hole; with only the other one open, a second patch of light. Adding the two effects together would give a larger patch of light. But interference means that when light passes through both holes simultaneously the pattern on the screen is more complicated – not least because the brightest part of the pattern, it turns out, is at a spot on the screen halfway between the two bright spots you would get if the holes were open separately, exactly where you might expect there to be a dark shadow.

So far, so good. Light *is* a wave. Unfortunately for this simple picture, though, there is also very good evidence that light is made of particles, called photons. And the way particles pass through two holes in a wall is very different, according to our everyday experience, from the way waves pass through holes in walls.

Suppose that the two holes were indeed holes in a wall, and that you stood on one side of the wall with a large pile of rocks, throwing the rocks, one at a time, in the general direction of the wall, but without bothering to aim very precisely. Some of the rocks would go through one hole and some through the other, building up two piles behind the wall. The pattern (two piles of rocks) would be exactly the same as the pattern you would get if you blocked off one hole for half the time and the other hole for the other half of the time. You certainly do not get a pile of rocks centred halfway between the two holes, right behind the solid part of the wall. Particles going through holes one at a time do *not* interfere with one another.

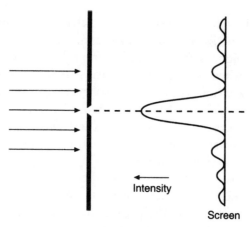

Figure 2
An electron beam passing through a single hole produces a distribution with most of the electrons in line with the hole. This is the way a beam of particles ought to behave.

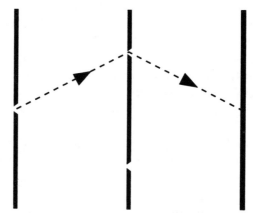

Figure 3
An electron or a photon passing through one of a pair of holes ought to behave as if it were passing through a single hole, if common sense is any guide. According to common sense, the presence of the second hole should have no effect.

Of course, if many particles are going through the holes at the same time, then it is easy to see that they might interfere with each other, jostling one another to make a different kind of pattern on the other side. After all, we are used to the idea that water itself is made of particles – molecules of water – and that doesn't stop ripples on a pond forming well-behaved waves. It is possible to imagine that in the same way a flood of photons from a lamp could act together like a wave when passing through the two holes. But the mystery deepens when we look at what happens to *single* photons going through the experiment with two holes one at a time.

It is important to stress that this experiment has actually been done, by a team working in Paris in the mid-1980s. They have, in effect, watched single photons going through the experiment with two holes and interfering with themselves. When I wrote *In Search of Schrödinger's Cat*, the evidence for the behaviour of light under these circumstances was very strong, but still, strictly speaking, circumstantial. Now, we *know* what happens, beyond a shadow of a doubt, when a single photon passes through the experiment.

All that we actually see, of course, is the pattern made by the light on the screen after it has passed through the two holes. Imagine that the light source is turned down so faint that only a single photon at a time emerges and passes through the experiment (this, in effect, is what physicists can now do, although the trick requires great skill and sophisticated apparatus). Now imagine that the screen on the other side of the two holes is a photographic plate which records the arrival of each photon as a white dot. As the individual photons pass through the experiment, in each case you see exactly what you would expect – a single photon leaves the light source, and makes a single white spot on the photographic plate. But as first hundreds, and then thousands, then millions of photons pass through the experiment, you behold a fantastic sight. The individual white dots on the photographic film congregate in exactly the bright stripes of the typical wave-interference pattern, leaving dark stripes in between.

Although each photon starts out as a particle, and arrives as a particle, it seems to have gone through both holes at once, interfered with itself, and worked out just where to place itself on the film to make its own minute contribution to the overall interference pattern. This behaviour encompasses two mysteries. First, how does the single photon go through both holes at once? Secondly, even if it does perform this trick, how does it 'know' where to place itself in the overall pattern? Why doesn't every photon follow the same trajectory and end up in the same spot on the other side?

Now, mysterious though all this is, you might argue that there is something odd about light. Indeed there is. Light (strictly speaking, electromagnetic radiation) always travels at the same speed, the speed of light (denoted by c). However you move, and however the light

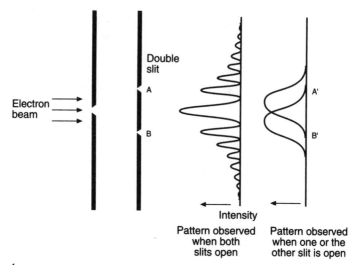

Figure 4

But both electrons and photons behave as if they know about the presence of the other hole. The pattern we see when both holes are open is not the same as the pattern you would get by adding together the patterns you get when each of the holes is open on its own. Does this mean that electrons are really waves?

source moves, when you measure the speed of light you always get the same answer. This has deep implications, as we shall see when I discuss the theory of relativity; it certainly is not like the behaviour of anything else in the everyday world. In addition, photons have no mass, another strange and non-commonsensical property. Perhaps the weird behaviour of photons passing through the experiment with two holes is due to the fact that they are weightless and travel at the speed of light? Or perhaps it is just one more weird property of light to add to the list. As Ralph Baierlein has put it, 'light travels as a wave but departs and arrives as a particle'.[1] Maybe that is just one of those things, a special property of light?

Unfortunately, this is not the case. You can do exactly the same trick with electrons – which, although not exactly the kind of particle we are used to handling individually in our everyday lives, have not only mass but also electric charge, and have the decency to move at different speeds depending on circumstances. Yet electrons also travel as waves but depart and arrive as particles. And that is much harder to dismiss as just one of those things.

[1] Baierlein, *Newton to Einstein*, p. 170.

ELECTRONIC INTERFERENCE

Electrons are very much part of the particle world. They were first identified as particles by J. J. Thomson, working at the Cavendish Laboratory in Cambridge, in 1897. Thomson showed that electrons are pieces that have escaped, or been broken off, from atoms – the first proof that the atom is not indivisible. Each electron has exactly the same mass (a little over 9×10^{-31} kg, which means 'a decimal point followed by 30 zeroes and a 9' kg). Each electron has the same electric charge (1.6×10^{-19} coulomb). They can be manipulated using electric and magnetic fields, and move faster or slower according to the way they are pushed and pulled around. In very many ways, electrons behave like tiny, electrically charged bullets.

And yet, by the late 1920s, 30 years after the discovery of electrons, it was clear that electrons also behave as waves. One of the people who proved this, in 1927, was George Thomson, the son of J. J. The evidence for the dual nature of electrons, the so-called wave–particle duality, had been well-established long before the middle of the 1980s. But it was only in 1987 that a Japanese team actually carried out the experiment with two holes using electrons.

Before that date, textbooks (including Feynman's) and popular accounts (including mine) had described such experiments, confidently assuring readers that, although these were just 'thought experiments', on the basis of everything known about electrons, it was possible to say how they would behave when confronted by two tiny holes in a wall. It was, however, 90 years after electrons were identified as particles, and 60 years after they were identified as waves that a team from the Hitachi research labs and Gakushuin University in Tokyo actually did the double-slit trick for electrons.

The 'double slit' in their experiment was formed from an instrument known as an electron biprism, and the screen the electrons arrived at on the other side was, in effect, a TV screen, on which the arrival of each electron made a spot of light which stayed on the screen. So the arrival of successive electrons gradually built up a pattern on the screen.

The results of the experiment were exactly the same as those for the equivalent experiment involving photons. The source of the electrons was the tip of an electron microscope, a standard and well-understood piece of equipment. The electrons emerged from the tip of the electron 'gun' as particles; they arrived at the screen on the other side as particles, each making a single spot of light. But the pattern that built up on the screen was an interference pattern, showing that the electrons had travelled through the two holes as waves.

You might still wish to quibble about this strange behaviour of electrons. After all, you cannot hold a single electron in your hand. Nobody has seen an electron, only the spots they make when they strike suitably sensitive screens. And we know from everyday experience that these bizarre interference effects do not occur when we throw rocks through holes. Neither rocks, nor baseballs, nor anything else in the everyday world shows this strange wave–particle duality.

Well, the physicists have an answer to that, too. If you want proof that particles big enough to be seen also behave like waves when they pass through the experiment with two holes, they have it.

The particles in question are atoms. Admittedly, you still cannot see atoms with your own eyes, or hold a single atom in the palm of your hand. But individual atoms, cupped in magnetic fields, can now be photographed. The achievement (described by, for example, Hans von Baeyer, in *Taming the Atom*) is all the more remarkable because the concept of atoms was only fully accepted by scientists at the beginning of the twentieth century. Indeed, Albert Einstein got his PhD for work establishing, among other things, the reality of atoms. Although atoms are much bigger than electrons, they are still tiny by everyday standards. An atom of carbon, for example, weighs in at just under 2×10^{-26} kg, 22 million times the mass of an electron. The size of an atom is about one ten-millionth of a millimetre, which means that it would take ten million atoms to stretch across the width of single serration on the edge of a postage stamp. But individual atoms have, nevertheless, been photographed, and their images can even be displayed on TV screens in 'real time'.

The experiment with two holes was first performed using atoms only at the beginning of the 1990s. A team at the University of Konstanz, in Germany, used atoms of helium passing through slits 1 micrometre (one millionth of a metre) wide in gold foil, with a detector on the other side. This time, the build-up of the interference pattern could not be displayed directly on a TV screen, but the measurements of the number of helium atoms arriving at different parts of the detector 'screen' showed the now-familiar pattern. Atoms, too, travel as waves but arrive as particles.

Several other groups of researchers announced similar results in the early 1990s. One, at MIT, involved a beam of sodium atoms. In all these experiments, the results are the same. A single atom passing through the experiment with two holes goes both ways at once, and interferes with itself. An atom, it seems, can be in two places (both holes) at the same time.

In one final (for now) twist on the theme, researchers at the US National Institute of Standards and Technology, in Boulder, Colorado, and the University of Texas reported in 1993 that they had turned this kind of experiment on its head. Instead of sending atoms through a two-hole experiment, they had trapped pairs of atoms in a magnetic field, and, in effect, used the atoms as the 'holes', bouncing light off them and measuring the interference patterns produced. The waves bouncing off the atoms spread out in much the same way as waves spreading out from the holes in the two-hole experiment. This new experiment only works, of course, because the atoms are particles, which can be trapped by magnetic fields and do the scattering. There is no neater example of wave–particle duality than this combination of experiments involving atoms – particles big enough to be photographed, remember – and interference.

Since these strange effects do not show up for rocks, or baseballs, or anything else we can handle and touch and see with our own eyes, there must be some level at which the rules of the quantum world cease to apply. Somewhere on the scale of sizes between an atom and a human being the quantum rules cut out, and the rules of classical physics cut in. Just where that level is, and why the changeover occurs, are themes that will be addressed later in this book. The answers strike at the heart of our concept of reality.

But for now the important point to stress, again and again, is that all these experiments have now been done. The results came as no surprise to physicists. Any competent physicist since 1930 could have predicted, using quantum theory, how the experiments would turn out. But they *might* have turned out differently – quantum theory *might* have been wrong. But no. Right down at the very deepest level, at the heart of the mystery, when the key experiments were carried out at the end of the 1980s and in the early 1990s they came up with 'answers' exactly in line with quantum physics. So how does quantum physics account for this peculiar behaviour?

THE STANDARD VIEW

The standard interpretation of what is going on in the quantum world is known as the Copenhagen Interpretation, because it was largely developed by the Danish physicist Niels Bohr, who worked in Copenhagen. Other people, notably including the Germans Werner Heisenberg and Max Born, made major contributions to the package of ideas that became the Copenhagen Interpretation, but Bohr was always its most evangelical proponent. The package was essentially complete by 1930, less than a human lifetime ago. Since then, it has been the basis of virtually all practical work involving the quantum world, and it is the story taught to aspiring physicists in university and college. But it rests upon some quite bizarre concepts.

The key concept is the so-called 'collapse of the wave function'. In seeking to explain how an entity such as a photon or an electron could 'travel as a wave but arrive as a particle', Bohr and his colleagues said that it was the act of observing the wave that made it 'collapse' to become a particle. We can see this at work in the electron version of the experiment with two holes – the electron passes through the experiment as a wave, then 'collapses' into a single point on the detector screen.

But this is only part of the story. How can the wave of a single electron interfere with itself, and how does it choose which point on the screen to collapse onto? According to the Copenhagen Interpretation, this is because what actually passes through the experiment is a wave of probability, not a material wave at all. The equation that describes how a quantum wave moves – the wave equation derived by the Austrian Erwin Schrödinger – is not describing a material wave like the ripples on a pond, but is actually describing the probability of finding the photon (or electron, or whatever) at a particular place.

On this picture, largely derived from Born's work, an electron that is not being observed literally does not exist in the form of a particle at all. There is a certain probability that you might find the electron here, and another probability that you might find it there, but in principle it could turn up literally anywhere in the Universe. Some locations are very probable – in the bright fringes of the two hole experiment – and some are extremely unlikely – in the dark fringes. But it is actually *possible*, although highly unlikely, that the electron

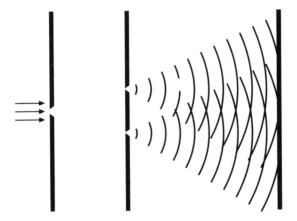

Figure 5
The standard explanation of the puzzle posed by Figure 4 (p. 6) is that 'probability waves' pass through both holes and decide where each particle in the beam ends up. Probability waves interfere in just the same way that water waves do.

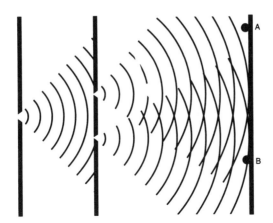

Figure 6
But when we look for particles, we find particles (A and B in this example)! The probability waves decide where the particles are, but we never see the waves. We don't really know what is travelling through the experiment. This strange behaviour has given rise to the quip that an electron (or a photon) 'travels as a wave but arrives as a particle'.

might turn up on Mars, or in the TV set of the man next door, instead of in the interference pattern at all.

Once the electron is observed, though, the odds change. The wave function collapses (maybe on Mars, if somebody is looking there, or more probably in the interference pattern), and at that moment it is 100 per cent certain where the electron is. But as soon as you stop looking, the probability starts leaking outward from that location. The probability of finding the electron in the same place that you last looked decreases, and the probability of finding it somewhere else increases as the probability wave spreads through the Universe.

Strange though it sounds, this is a very useful concept in practice, because in all practical applications, such as making TV sets and computer chips, we are dealing with huge numbers of electrons. If they all obey the strict rules of probability and statistics, that means that the behaviour of the bulk of electrons is predictable. If we know that 30 per cent of the electrons will go one way through a computer circuit, and 70 per cent of them will go the other way, we don't have to worry about which way an individual electron goes. In the same way, the owners of a casino know that the rules of probability will ensure them a profit in the long term, even if the occasional player has a big win at roulette. But Albert Einstein was so disgusted by the whole notion that he made his famous remark, 'I cannot believe that God plays dice with the Universe'; and the implications are laid bare when we get down to experiments involving single electrons, or single photons.

One piece of laying bare can be seen by thinking once again in terms of the experiment with two holes. This version of the experiment has not yet been carried out with single electrons, but slightly more complicated experiments have confirmed the way in which electrons behave, and there is no doubt that this is what would happen if you could carry out the experiment in this pure form.

First, remember what happens to the interference pattern (produced either by light or by electrons) when one of the holes is closed. The pattern disappears. Obviously, when only one hole is open the electrons must travel through that hole, and that hole alone, to reach the detector screen. But if you think of electrons simply as particles, that is strange enough in itself. How does an electron passing through one hole 'know' whether or not the other hole is open? A simple particle, travelling through one hole of the experiment with two holes, would neither know nor care whether the other hole was open or not. But even if you set up the experiment so that the second hole is closed (or open) when each electron leaves the 'gun', but is then opened (or closed) before the electron reaches the first hole, it will 'choose' the appropriate path to the target screen to make the right kind of overall pattern. You can even set up the experiment to open and close the second hole at random. Each electron chooses a trajectory at one hole which depends on whether or not the other hole is open *at the same time*.

It seems that the electrons are aware of more of the world than their immediate locality. They are aware of conditions not just at one hole, but throughout the entire experiment. This non-locality is a fundamental part of quantum mechanics, and worried Einstein deeply; it is the origin of his reference to 'spooky action at a distance', although when he made that remark he was thinking about an even stranger manifestation of non-locality, which I shall describe shortly.

So far, though, all the evidence has come simply from looking at the pattern made on the detector screen while trying different combinations of open and closed holes. Why not look to find out what is happening at the holes themselves? Imagine setting up a pair of detectors, one alongside each of the two holes in the experiment, and sending the electrons through, one at a time. Now, you can look to see whether the electron passes through both holes at once, as a wave, or whether it passes through just one of the holes (or, indeed, whether a half-electron passes each hole). And you can also keep an eye on the detector screen, to see what kind of pattern builds up there after many electrons have gone through the experiment. What you find in such a situation is that each electron is always seen to be a particle, travelling through one hole or the other. It behaves like a little bullet. And, lo and behold, the interference pattern disappears. Instead, the pattern on the screen becomes the pattern produced by little bullets travelling through each hole independently (or by rocks thrown through two holes in a wall). The act of observing the electron wave makes it collapse and behave like a particle at the crucial moment when it is going through the hole. Don't imagine, though, that we have escaped from the puzzle of non-locality. In fact, we only need to look at one of the two holes to change the pattern. If we do, we see only bullet-like electrons going through, and we see the pattern appropriate to particles on the screen. Somehow, the electrons going through the second hole 'know' that we are looking at the other hole, and also behave like particles as a result.

And the probability aspect of the Copenhagen Interpretation still comes into the story. Assuming that the experiment has been set up perfectly symmetrically, you will find that exactly half of the electrons follow each possible route. 50 per cent go through one hole, and 50 per cent go through the other hole. There is no way to predict in advance which hole an individual electron will go through, and therefore which blob it will arrive at on the detector screen. Like tossing a coin and getting a sequence of heads, just by chance several electrons in a row may go through the same hole. But after a million electrons have gone through the two holes while being watched, there will be half a million in one blob on the screen and half a million in the other blob. The probability wave is still doing its work, even when you are watching the electrons and know that they are behaving as particles.

Bohr argued that what matters is not the behaviour of a single electron, or even the behaviour of a million electrons. What matters is the whole experimental set-up, including the electrons, the two holes, the detector screen *and the human observer*. It is impossible to say that an electron 'is' a wave or 'is' a particle. All that can be said is that if an experiment is set up in a certain way, and certain measurements are made, then you will see certain results. Set the experiment up to measure waves, and you will get an interference pattern; set the experiment up to monitor particles passing through the holes, and you will see particles passing through the holes. You can even wait until after the electrons have left the 'gun' before deciding whether or not to switch on the detectors at the two holes and look for particles; in every case, the ultimate outcome (the pattern on the screen) depends on the whole experimental set-up. And this holistic view of the quantum world leads us into very deep philosophical water.

DEEP WATER

The Copenhagen Interpretation held sway for more than 50 years, from 1930 until well into the 1980s, almost unopposed by the vast majority of physicists. They did not care about the deep philosophical puzzles associated with the Copenhagen Interpretation – indeed, many still do not care – provided that it could be used as a practical tool for predicting the outcome of experiments. But in recent years there has been growing unease about what quantum theory 'means', and increasing efforts have been made to find alternative interpretations.

The main problem is with this business of the collapse of the wave function. It is all very well Bohr telling us that the whole experiment has to be considered, and that the way the waves will collapse will depend on the overall experimental setup; but there is no such thing as a pure, self-contained experiment. This interpretation of quantum theory is telling us that entities such as electrons are only real in so far as they are observed – that the measuring apparatus is, in some sense, 'more real' than the photons and electrons and all the rest. This is not my interpretation of the Copenhagen Interpretation; it is what Bohr and Heisenberg and their colleagues explicitly stated. For example, Heisenberg said: 'The Copenhagen Interpretation regards things and processes which are describable in terms of classical concepts, i.e. the actual, as the foundation of any physical interpretation.'[1] In other words, the atoms of which everything in the classical world is made are somehow less real than the things atoms are made into! This struck many people as downright weird even in the 1930s; it is even harder to swallow now that atoms have been photographed.

Applying this line of reasoning – the Copenhagen Interpretation – to the experiment with two holes, somebody has to look at the experiment to make it settle down into one state. In the words of Heinz Pagels, who was at the time (1981) President of the New York Academy of Sciences, and certainly understood what quantum theory was all about: 'There is no meaning to the objective existence of an electron at some point in space, for example at one of the two holes, independent of actual observation. The electron seems to spring into existence as a real object only when we observe it!'[2] But the experimenter is also part of the outside world, not just a piece of the experiment. People are made of, among other things, electrons; what collapses their wave functions to make them behave as localized objects within the experimenter's body? Presumably, interactions with the world outside the observer. And what makes the world outside the observer 'real', in this sense? More interactions, with more things (and observers), on a larger and larger scale. Take the Copenhagen Interpretation literally, and it tells you that an electron wave collapses to make a point on a detector screen because the entire Universe is looking at it. This is strange enough; but some cosmologists (among

---

[1] Quoted by Nick Herbert in Paul Davies (ed.), *The New Physics* (Cambridge: Cambridge University Press, 1989), p. 143.
[2] Pagels, *The Cosmic Code*, p. 144.

them Stephen Hawking) worry that it implies that there must actually be something 'outside the Universe' to look at the Universe as a whole and collapse its overall wave function.[1] Alternatively, John Wheeler has argued that it is only the presence of conscious observers, in the form of ourselves, that has collapsed the wave function and made the Universe exist. On this picture, everything in the Universe only exists because *we* are looking at it. I'll be looking more closely at such desperate remedies and counsels of despair later; but the fact that such arguments are seriously put forward by respected scientists is enough to demonstrate what deep water we are in.

Another problem concerns the relationship between the particle and wave aspects of a quantum entity's behaviour. Bohr described these as complementary properties, in the same way that the head and tail of a coin are complementary. If you place a coin flat on the table, it will have either the head uppermost or the tail uppermost, but not both at the same time. In the Copenhagen Interpretation, an entity such as an electron is neither a wave nor a particle, but something different, something we cannot describe in everyday language. But it will show us either a particle face or a wave face, depending on which measurements we choose to carry out on it – which way up we choose to place the quantum coin. Indeed, it may have other properties as well, that we are not clever enough to measure at all, and know nothing about.

This complementarity, or wave–particle duality, is related to the famous uncertainty principle discovered by Heisenberg. The simplest version of this principle tells us that it is impossible to measure both the position and the momentum of a quantum object at the same time. Momentum is simply a measure of where such an object is going, and how fast. It is, in many ways, a wave property – waves must be moving somewhere, or they would not be waves. Position is a definite particle property – a wave is by nature spread out, whereas a particle is confined in one place. We can make measurements which observe the position of an electron, or we can make measurements which tell us which way it is moving, and in either case we can make the measurements as accurate as we like. But trying to measure the position very accurately blurs the electron's momentum, by a quantifiable amount, and vice versa.

---

[1] See, for example, Hawking's book *A Brief History of Time*, or my own *In Search of the Big Bang*.

This is not, as some textbooks still mistakenly suggest, solely a result of the practical difficulty of making measurements. It is not simply because in measuring the position of the electron (perhaps by bouncing photons off it) we give it a kick, which changes its momentum. A quantum object *does not have* a precisely defined momentum and a precisely defined position. The electron itself does not 'know' within certain limits where it is or where it is going. Exaggerating only slightly, if it knows exactly where it is, it doesn't know where it is going at all; if it knows exactly where it is going, it doesn't have the faintest idea where it is. Usually, though, a quantum object has an approximate idea of both where it is and where it is going. But the important word here is 'approximate'; hard though it is to understand from the 'common-sense' viewpoint of our everyday world, the quantum entity cannot be pinned down to a definite location, and there is always some uncertainty about where it is going.

This is crucially important, for example, in nuclear fusion reactions, where the quantum uncertainty allows atomic nuclei that are not close enough to touch one another, according to the ideas of classical physics, to overlap with one another and combine. Some of these nuclear reactions are what keeps stars hot. Without quantum uncertainty, the Sun would not shine the way it does.[1]

These are difficult ideas to come to terms with, but I am not going to lead you through the history of how they came to be developed, or the evidence that this is indeed the way the quantum world works. Many other books, including my own, are now available to fill in those details. I shall be more concerned, in this book, with where the Copenhagen Interpretation breaks down, and what might supersede it. Uncertainty, though, really does seem to be a fact of life at the quantum level. It does not show up in the everyday world for the same reason that the wave–particle duality does not show up in the everyday world. The equations which describe these phenomena contain a number known as Planck's Constant, after the quantum pioneer Max Planck. Planck's Constant is tiny compared with the masses and momenta that apply to everyday objects. It has a value of just 6.55×10^{-27} erg seconds (don't worry about the units; all that matters is that the masses are measured in the equivalent units, grams). Quantum effects only become dominant for objects with masses in the same ball park – like the electron, with its mass of 9×10^{-31} kg, or,

---

[1] See my book *Blinded by the Light*.

to use the more directly comparable units to go with those erg seconds, 9×10^{-28} g. Start dealing in masses much bigger than those of atoms, and the quantum effects are so small that their influence can be ignored – *except* for the fact that everything larger than atoms is itself made up of atoms.

It's worth taking a breath here and trying to get a feel for just how far away from our everyday experience the quantum world is. The number 10^{-27} means one billionth of a billionth of a billionth. If an object was 10^{-27} of a centimetre across, it would take a billion billion billion of those objects to stretch across a distance of one centimetre. So how big a distance would we cover if we took a billion billion billion objects, each 1 cm across – sugar cubes, say – and laid them end to end? The answer is 10^{27} cm. How big is that? Well, the standard unit of length measurement in astronomy, the distance light can travel in one year (one light year) is about 10^{18} cm; so 10^{27} sugar cubes side by side would stretch across a distance of a billion (10^9) light years. The most distant objects known in the Universe, some quasars, are about 10 billion light years away. So 10^{27} sugar cubes would stretch one-tenth of the way to the most distant known quasar. In very round terms, the quantum world operates on a scale as much smaller than a sugar cube as a sugar cube is compared with the entire observable Universe. To put it another way, people are about midway in size, on this logarithmic scale, between the quantum world and the whole Universe – and we claim to be able to understand what is going on at both extremes.

We don't expect wave–particle duality to show up for a brick, or a house, or a person, because those things are so large compared with Planck's Constant. Physicists do, now, expect wave–particle duality to show up for quantum objects, although one of the key features of the Copenhagen Interpretation has been that you cannot see both aspects at once. Bohr was quite explicit about this, claiming that it is impossible in principle to see an entity such as a photon or an electron simultaneously exhibiting wave-like properties and particle-like properties. Unfortunately for Bohr, and for the Copenhagen Interpretation, experimenters are now challenging even that claim, as we shall see later.

The bottom line is that the Copenhagen Interpretation works, in the sense that it provides a series of recipes – involving uncertainty, the collapse of the wave function, probability, the role of the observer and the holism of experiments – which physicists can use to predict the outcome of experiments. But it doesn't *explain* anything. This realization is not new. Einstein spent ten years of his life fighting a friendly running battle in correspondence with Bohr, trying to show up the failings and absurdity of the Copenhagen Interpretation. And the best-known example of quantum absurdity was also developed, by Schrödinger, in an attempt to persuade his colleagues that the whole package of ideas was so ridiculous that it ought to be abandoned. I refer, of course, to the famous cat-in-the-box 'thought experiment', which for all its familiarity (the cat was 60 years old in 1995) is still worth summarizing as an example of the difficulties that any improved interpretation of quantum theory – any interpretation which really does *explain* things – must be able to explain.

THE CAT IN THE BOX

One of the strangest features of the Copenhagen Interpretation, brought out most clearly by the cat-in-the-box 'experiment', is the role of a conscious observer in determining what happens in the microworld. The simplest example of this is to imagine a box which contains a single electron. If nobody looks in the box, then according to the Copenhagen Interpretation there is an equal probability of finding the electron anywhere inside the box – the probability wave associated with the electron fills the box uniformly. Now imagine that, still without anyone looking inside the box, a partition is automatically lowered in the middle of the box, dividing it into two equal boxes. Common sense tells us that the electron must be in one side of the box or the other. But the Copenhagen Interpretation tells us that the probability wave is still evenly distributed across both half-boxes. That means that there is still a 50:50 chance of finding the electron in *either* side of the box. The wave only collapses, with the electron becoming 'real', when somebody looks into the boxes and notices which side of the partition the electron is. At that moment, the probability wave on the other side of the partition vanishes. If you close the box up again, and stop looking at the electron, its probability wave will spread out once more to fill the half-box in which the electron has been located, but it will not spread back into the other half of the box.[1]

The physicist Paul Davies has summed up the situation succinctly: 'It is as though, prior to the observation, there are two nebulous electron 'ghosts' each inhabiting one chamber, waiting for an observation to turn one of them into a 'real' electron, and simultaneously to cause the other to vanish completely.'[2] That word 'simultaneously' is also important here, pinpointing that this is another example of non-locality at work. But before I get on to the implications of that, I want to explain Schrödinger's demonstration of the absurdity of the claim that the observer is responsible for the reality of the electron existing in one half of the box or the other.

Schrödinger's puzzle first appeared in print in 1935. It depends upon setting up a quantum situation where there is a precise 50:50 chance of either of two outcomes. As it happens, he used radioactive decay in his original example, because radioactive sources also obey the rules of probability; but we can easily recast this in terms of the electron in the divided box. Schrödinger himself also referred to the experiment being carried out in a steel chamber; this has come down in quantum folklore as a 'box' containing, among other things, the cat in question. I prefer to interpret the term 'chamber' more generously, giving the cat room to enjoy life while it can. But none of this affects the thrust of Schrödinger's argument.

So imagine that the whole system I have already described – the double box, single electron and automatic sliding partition – is sitting on a table in a closed, windowless room. The partition has slid across, dividing the box into two halves, and there is a precise 50:50 chance that the electron is in either half of the box. Outside the box, there is an electron detector, which is wired up to an apparatus which will flood the room with poisonous gas if it detects an electron. In the corner of the room there is a cat, quietly minding her own business. Schrödinger described this set-up as a 'diabolical device',[3] but remember that this is only a 'thought experiment', and no real cat has ever suffered the indignities I am about to describe.

Schrödinger asked us, in effect, to imagine what happens when one

---

[1] At least, not with an equal probability; there will be a tiny (*very* tiny) probability that the electron might be in the other half of the box, or outside the box altogether, but that can be ignored for the purpose of this example.

[2] Davies and Brown, *The Ghost in the Atom*, p. 22.

[3] See Wheeler and Zurek, *Quantum Theory and Measurement*, p. 157.

half of the electron box is now opened up, automatically, allowing the electron, if it is in that half of the box, to escape. There has still been no human observation of what is going on in the locked room. According to the Copenhagen Interpretation, there is still a 50 per cent chance that the electron is in the sealed half of the box, but there is now also a 50 per cent chance that the electron is in the room at large. Since this is a thought experiment, we can imagine that the detector is so sensitive that it can reliably detect the presence of a single electron added to the contents of the room. If the electron has escaped from the box, it ought to be detected by the machine, which will trigger the release of the poison gas, killing the cat.

You might think that this is what happens, even when nobody is looking: either the electron escapes from the box, or it does not. If it does not, the cat is safe; if it does, the electron's probability wave will collapse when the detector 'notices' the electron, and the cat is doomed. But Bohr said that this common-sense view is wrong.

What the standard interpretation of quantum theory tells us is that because the electron detector is itself composed of microscopic entities of the quantum world (atoms, molecules and so on) and the interaction with the electron takes place at this level, the detector is also subject to the quantum rules, including the probability rules. On this picture, the wave function of the whole system does not collapse until a conscious observer (preferably equipped with a gas mask, if they want to be sure of staying conscious) opens the door to look inside. At that moment, and *only* at that moment, the electron 'decides' whether it is inside or outside the box, the detector 'decides' whether it has found an electron or not, and the cat 'decides' whether it is dead or alive. Until somebody looks inside the room, the Copenhagen Interpretation describes the situation inside as a 'superposition of states' – or, in Schrödinger's words, 'having in it the living and the dead cat (pardon the expression) mixed or smeared out in equal parts'.[1]

Depending on how you like to view the situation, you can imagine that the room contains a cat that is both dead and alive at the same time, or a cat that is neither dead nor alive, suspended in limbo. But you cannot, if the Copenhagen Interpretation is correct, imagine that the room contains either a simple dead cat or a simple live cat, until somebody looks.

[1] Wheeler and Zurek, *Quantum Theory and Measurement*, p. 157.

The whole point of this argument is to highlight the absurdity of the Copenhagen Interpretation, so don't be surprised if you can pick holes in it. The obvious puzzle is how you define a 'conscious' observer. Surely a cat is competent enough to know whether or not it has breathed in poison and died? Can't the cat's reaction to the events inside the room do the same job as a human observer looking in the door? But where do you draw the line? Working downwards from the human scale towards the quantum world, is an ant capable of collapsing the wave function? Or a bacterium?

Look at the puzzle the other way, from the quantum world upwards. It is all very well saying that the electron detector cannot make the wave function collapse because it is solely made up of quantum entities such as atoms and molecules – but a human being (or a cat) is also made up of atoms and molecules. If the detector is *not* competent to collapse the probability wave, why should we be? And is life a requirement of a conscious observer, in this sense of the term? Would a sophisticated computer be able to collapse the wave function by looking into the room?

Moving even further out from the original electron, what is the situation if the person who looks into the room, to see whether or not the cat is dead, is alone in the building, which is locked up for the night? The strict Copenhagen Interpretation says that the superposition of states (Schrödinger's smearing out) embraces this observer, too, until somebody else, outside the building, looks in to see how the experiment is going (or, perhaps, telephones to ask what the state of play is). Not just the cat, but the human observer also can be in limbo until somebody looks. And who looks at the person outside the building, to collapse *their* wave function? Shouldn't the whole process go on indefinitely, in an infinite regress?

The crucial question is where you draw the line between quantum probabilities and what we think of as reality. How many molecules, if you like, must a system contain before it becomes 'real' and can collapse wave functions? And how must those molecules be arranged in order for the system to do the trick?

That is the kind of puzzle that is now taxing philosophers and quantum mechanics. They all know that quantum theory *works*; but they want to know *why* it works, and to have some understandable image of what is going on inside a closed room when nobody is looking. But there is more to the puzzle than even the simple version of the cat-in-the-box scenario suggests. Before I move on to the meaning of quantum mechanics, I want to lay bare the deeper aspects of the mystery, with the aid of the daughters of Schrödinger's cat.

ANOTHER ASPECT OF REALITY

It is a sign of how dramatic progress in physics has been that although nobody has really tried locking up a cat in this way to see what happens to it, another thought experiment, which was dreamed up by Albert Einstein just before Schrödinger thought up his cat-in-the-box puzzle, actually became practical reality in the 1980s. But it may be just as well that Einstein never lived to see this thought experiment turned into practical reality, because although, like the cat-in-the-box puzzle, it was designed to highlight the absurdity of quantum theory, when the test was actually carried out quantum theory passed with flying colours.

Einstein did not work this particular idea out on his own. He developed it with Boris Podolsky and Nathan Rosen shortly after he moved to Princeton in the early 1930s, and the puzzle appeared in print under their joint names in 1935 – the same year that Schrödinger published the cat-in-the-box 'paradox'. It is known as the 'EPR paradox', because it highlights the non-logical (by everyday common-sense standards) nature of quantum reality.

The puzzle was refined by David Bohm, an American physicist who settled in England, in 1951; but it remained purely a thought experiment at that time. In the mid-1960s, however, John Bell, an Irish physicist who worked at CERN, in Geneva, found a way to express the puzzle in terms of an experiment that could, in principle, be carried out on pairs of photons emitted from an atom simultaneously in two different directions. At that time, not even Bell thought that the experiment was a practical possibility. But over the next 20 years or so several researchers took up the challenge of measuring the relationships described by Bell. The most comprehensive and conclusive of these experiments were carried out by Alain Aspect and his colleagues, working at Orsay, in Paris, in the early 1980s. They demonstrated beyond reasonable doubt that common sense (and Einstein) is wrong, and that non-locality really does rule in the quantum world. It is Bell's version of the EPR paradox, as tested by Aspect, that I will describe here.

The property of photons that is measured in the Aspect experiment is called polarization. One way of thinking of polarization is that each photon of polarized light carries an arrow pointing in a certain direction – up, down, sideways, or somewhere in between. There are many strange features of the behaviour of polarized light, some of which I discuss in Chapter Three, but all that matters here is that it is possible to measure these different facets of the photon polarization, and that these properties are correlated in accordance with the quantum rules. Simplifying the actual situation slightly, it may be true that one photon must point upwards, and the other must point sideways, but there is nothing in the rules to say which photon points which way. When the two photons are emitted from the atom, they exist, like Schrödinger's cat, in a superposition of states, until somebody measures the polarization of one of them. At that moment, the wave function of that photon collapses into one of the possible states – pointing up, perhaps. At the same moment, the wave function of the *other* photon must also collapse, into the other state – in this case, pointing sideways. Nobody has looked at the other photon, and by the time the measurement is made the two photons may be far apart (in principle, on opposite sides of the Universe); but when one wave function collapses so does the other one. This is what Einstein referred to as 'spooky action at a distance'; it is as if the two quantum entities (two photons, in this case) remain tangled up with one another for ever, so that when one is prodded the other one twitches, instantaneously, no matter how far apart they are.

This was particularly abhorrent to Einstein because, as we shall see, his theory of relativity is based on the fact that light always travels at the same speed, and nothing can be accelerated from travelling slower than light to travel faster than light. According to relativity theory, at least as it was originally understood, nothing can link two particles instantaneously across space. As we shall see, there may be more to relativity theory than even Einstein realized; but at the time this was, especially for him, a powerful argument against the possibility of such action at a distance.

But how can evidence for (or against) spooky action at a distance actually be obtained by experiments? It is no good measuring both photons; you will always get the right answers (one up, one sideways, in this example) but you will never 'see' the instantaneous connection at work. For all you could tell, by making those measurements, the properties of each photon might have been determined at the moment they left the atom, as common sense would suggest. The trick of catching action at a distance – non-locality – at work is to work with three connected measurements (three angles of polarization, in the Aspect experiment), but actually to measure only two of them, one for each photon.

Because polarization is an unfamiliar property, it may help to think of what is going on in terms of colours, with the caution that Aspect's team was not actually measuring colours in this way. Suppose that the atom actually emits not pairs of photons, but pairs of coloured particles, like tiny snooker balls. The colour of each ball may be red, yellow, or blue, say; but in each pair of balls the two colours will be different.

Putting this back into quantum language, when the atom shoots out two balls in opposite directions, the Copenhagen Interpretation tells us that neither of the balls has a definite colour. Both exist in a superposition of the three possible states. When the experimenter 'looks' at one ball, its wave function collapses, and it takes on a definite colour. At the same moment, the wave function of the other ball collapses, and it takes on one of the other two possible colours – but we do not know, from our single measurement alone, which one.

Now, it is possible to make a measurement of one ball which tells us whether or not the ball is blue. The answer to that question will give information about the state the other ball has collapsed into, but not complete information about the state of the other ball. Suppose that the result of our measurement is 'blue'. Then the state of the other ball must be either 'red' or 'yellow'. The only other possible result of our measurement is that the state of our ball is 'not blue'. In that case, since we have not specified whether our ball is actually red or yellow, the other ball may have taken on any of the three possible colours, but it is more likely to be blue than to be red or yellow, for the following reason.

If the first ball is 'blue', then the second ball must be either 'red' or 'yellow'. So there is a 50:50 chance of finding it in either of these two colour 'states'. If the first ball is 'not blue', however, there are two different possibilities for its own state. First, it might be 'red'. If so, then the second ball might be either 'blue' or 'yellow'. Second, the first ball might be 'yellow'. If so, the second ball may be either 'blue' or 'red'. So there are now *four* possible alternatives for the second ball. Two of these alternatives are *both* 'blue', so there is a 50 per cent (2 in 4) chance that the ball is 'blue'. One of the four possibilities is 'red', and one of the four is 'yellow'. So there is a 25 per cent (1 in

4) chance of the ball being 'red', and a 25 per cent chance of the ball being 'yellow'. Of course, it must be one of the three colours, once it has been looked at, and, sure enough, the percentages add up to 100 per cent.

The act of measuring the state of the first ball changes the odds on finding a particular colour when we measure the state of the second ball. To see how the odds change according to the way we measure the first ball, you have to make the measurements very many times, on very many balls, just as you need to toss a coin very many times to see the pattern of 50 per cent heads and 50 per cent tails emerging clearly. The crucial point, though, is that Bell showed that the statistical pattern that ought to emerge if non-locality is at work is different from the pattern that would emerge if each ball 'chooses' its colour when it leaves the atom and stays that colour ever afterwards.

In this terminology, the experiment consists of asking pairs of questions about the two photons together, along the lines of 'Is one photon blue or not, and is the other photon yellow or not?' We can also ask, 'Is one photon blue or not, and is the other photon red or not?' Carry out such measurements many times, on many pairs of particles, and you end up with a list of answers, specifying how often the particles pair up so that they are respectively 'blue and not red', how often they pair up 'not blue and not yellow', how often they are 'blue and not yellow' and so on. What Bell showed is that if you ask questions like this very many times, using very many pairs of photons, then there is a statistical pattern in the answers you get. We can find out how often the combination 'blue and not yellow' occurs, compared with the combination 'not blue and not red', and all other possible combinations. And, I stress, because the quantum entities do not decide which colour they are until you look at them, whereas common-sense particles have chosen their colours the moment they leave the atom, and stick with them ever afterwards, the statistical pattern is different in the quantum world from the common-sense world.

Bell showed that if common sense holds then one particular set of measurements – one pattern of behaviour, which we can call pattern A – must always occur *more often* than another specific set of measurements – a second pattern of behaviour, pattern B. Common-sense logic says that pattern A is more common than pattern B. But the Aspect experiment (and many other experiments along the same lines) shows that this inequality is violated. The number of times that pattern

A occurs is *measured* to be *less* than the number of times pattern B occurs.

Although couched in mathematical language, the argument is based on common-sense logic. For example, common-sense logic tells you that the number of teenagers in the world must be less than the number of female teenagers plus men of all ages. The results of the Aspect experiment are equivalent, in logical terms, to discovering that there are, actually, more teenagers in the world than there are teenage girls and all men (teenagers *and* adults) put together. Bell's inequality is violated, which means that non-locality is at work, and quantum theory has been vindicated – although we *still* don't know what all this *means*.

Bell himself regarded quantum theory as 'only a temporary expedient'[1] and always hoped that physicists would come up with a theory that could explain even these oddities in terms of a real world that exists even when we are not looking at it or measuring it. But although the outcome of the Aspect experiment was, in this sense, the opposite of what he had hoped for (not the opposite of what he *expected*, in the light of the previous triumphs of the quantum theory), he later told physicist Nick Herbert that he was 'delighted – in a region of woolliness and obscurity to have come upon something hard and clear', even if that something ran counter to common sense and his own prejudices.[2]

Translating the implications of the Aspect experiment back into a slightly simpler example, these discoveries imply that if the atom emits two particles in different directions, and the quantum rules require that one must be red and the other yellow, but the rules do not specify which one is which colour, then the particles each exist in a superposition of states until a conscious observer notices which colour one of them is. At that moment the wave function of that particle collapses one way, *and* the wave function of the other particle collapses into the alternative colour. And, once again, it is worth emphasizing that this is not simply some wild idea dreamed up by a crazy theorist, nor is it even merely a carefully worked-out thought experiment. This non-local behaviour has been *proved* to occur, by real experiments carried out with photons. By recasting the experiment slightly to involve a single electron and a pair of kittens, we can update Schrödinger's famous thought experiment to take account of Aspect's

[1] Davies and Brown, *The Ghost in the Atom*, p. 51.
[2] Letter to Herbert, quoted in Herbert, *Quantum Reality*, p. 212.

measurements of the violation of Bell's inequality, and see once and for all just what non-locality and action at a distance imply.

THE DAUGHTERS OF SCHRÖDINGER'S CAT

Now comes the crunch. Here is the basic problem in all its glory.

Imagine two kittens, the twin daughters of Schrödinger's cat, each living in a space capsule, attended by automated equipment and supplied with plenty of food. The two space capsules are linked by a narrow tube, opening into the capsules at each end. In the middle of the tube, there is a box which has an automated sliding partition across its middle, and which contains – you guessed! – a single electron. Each of the two space capsules contains the usual diabolical device which will kill its respective cat if an electron emerges from the tunnel into the capsule, and the electron box in the middle of the tube completely blocks the tube, so nothing can pass from one capsule to the other. But the ends of the box are now also sliding partitions.

Remember that as long as nobody looks the electron's probability wave fills the box uniformly. When the sliding partition in the middle of the box divides it into two halves, there is a 50 per cent probability that the electron is on one side of the partition, and a 50 per cent probability that it is on the other side of the partition. So when the two ends of the box slide away, the probability wave will spread out into each of the two space capsules evenly. If the linking tube is now automatically severed, exactly down the middle of the partition that divided the electron box, we will be left with two unconnected space capsules, each containing a cat being tended to automatically, a diabolical device that will kill the cat if it detects an electron, and a 50 per cent electron probability wave. Electron wave, diabolical device and cat are now *all* in a superposition of states.

Because this is only a thought experiment, we can equip our hypothetical space capsules with the very best in propulsion systems that the laws of physics allow – although, of course, we will not allow them to violate Einstein's theory of relativity by travelling faster than light. We also assume that the kittens come from a hardy and (diabolical devices permitting) long-lived line. Now, with the two capsules separated, automatic rockets can fire to propel the two craft in opposite directions through space. They travel for several years. Eventually, one of them reaches a distant planet, where there are conscious (intelligent) observers. The other capsule is by then more than a light year away, having been carried in the opposite direction by the super-efficient rockets.

Curious to find out what is inside the capsule, the intelligent observers open the hatch and take a peek. At that moment, the wave function of the contents of the capsule collapses. It 'decides' whether or not the original electron entered the capsule that is being studied. If it did, the cat dies – or rather, once the observation has been made the cat *was* dead all the time, from back when the electron was released from its box. At the same moment that the aliens observe the dead cat, the other cat is released from its superposition of states, and 'becomes' alive. Alternatively, of course, the aliens may open the capsule to find a live cat. In which case, their act of observation has consigned the other cat to its fate. It isn't so much that each cat is both dead and alive at the same time, but as if, during all the years in space, there was one dead cat and one live cat, but complete ambiguity about which capsule contained which cat. Or as if each capsule contains two ghosts, representing alternative versions of history, one of which fades away while the other becomes real at the moment of observation.

How you choose to interpret what is going on is largely up to you, as far as the Copenhagen Interpretation is concerned. There is *no* official 'interpretation' at this level – all the Copenhagen Interpretation cares about is that if you carry out the same experiment thousands of times on thousands of pairs of cats you will always find half of the cats landing on the alien planet to be dead and half to be alive, while their counterparts are always in the opposite state. The standard interpretation doesn't even have anything to say about the inference that as well as the non-local action at a distance which signals from one capsule to the other instantaneously at the moment when the wave functions collapse, from one point of view there is an element of time travel involved.

You could argue that the act of observation sends a signal not just across space but also echoing back across time, back to the moment when the electron was released, determining which capsule it went into. This is, in fact, no harder to swallow than the notion of instantaneous signalling across space, because one of the things that emerges from Einstein's theory of relativity is that *if* a signal could travel faster than light then it would also be travelling backwards in time (which, of course, is one reason why such 'superluminal' signalling is usually dismissed as impossible).

Accepting the possibility of signals that go backwards in time may seem extreme, but it would have the merit, if it could be incorporated into a comprehensive interpretation of the quantum world, of doing away with the ghostly superposition of states typified by the fate of Schrödinger's cat and her daughters. Bell himself once said that given the choice he would prefer to retain the notion of objective reality and throw away the idea that signals cannot travel faster than the speed of light.[1] But to understand why that is both a drastic choice and (perhaps) a tenable one, we need to know more about the nature of light, whose behaviour lies at the heart of physicists' understanding of both relativity theory and quantum theory.

If you are the kind of person who reads the last page of a mystery novel first, and if you think you already know about the standard interpretation of relativity theory and quantum physics, by all means take a peek at the Epilogue now. But if you do, promise to come back and read the rest of the book, because like all good mystery writers I have some tricks up my sleeve with which to entertain you between now and the denouement. Some of those tricks, like those of some of the best magicians, involve mirrors; and they all reflect the mysterious nature of light itself.

---

[1] Davies and Brown, *The Ghost in the Atom*, p. 50.

CHAPTER ONE

Ancient Light

In science, what you regard as ancient history depends on your point of view. Descriptions of the Universe and how it works – theories and mathematical models – that do not incorporate the ideas of quantum mechanics are often referred to as 'classical' theories. By this criterion, Isaac Newton was a classical scientist, every bit as much as Archimedes was. Indeed, by this definition even Einstein's two theories of relativity are classical theories. And yet, twentieth-century physics was founded on two pillars, the quantum theory *and* the theory of relativity. Both changed the way scientists viewed the world, and both emerged at the beginning of the twentieth century. So from another perspective the ancient history of science concerns everything before about 1900. It is in that sense that I use the term when describing the ancient history of the investigation of light – everything from the time of the Ancient Greeks to the work of James Clerk Maxwell which showed, in the nineteenth century, that light is a form of electromagnetic radiation.

Early philosophers thought that light originated in the eyes, reaching out like the beam of a lighthouse, or like the stick of a blind person, to 'feel' the nature of the world at large. Empedocles, who lived in the fifth century BC, and was the person responsible for the idea that everything is composed of the four 'elements' (earth, air, fire and water), described how Aphrodite had fashioned the human eye out of the four elements, held together by love. She kindled the fire of the eye at the hearth fire of the Universe, so that it would act like a lantern, transmitting the fire of the eye out into the world and making sight possible.[1]

Empedocles realized that there must be more to light than this, and that the darkness of night is caused by the body of the Earth getting in the way of light from the Sun. Epicurus, who lived in the third

[1] See Kathleen Freeman, *Ancilla*.

century BC, had similar views; his ideas were summarized by the Roman author Lucretius, who wrote *On the Nature of the Universe* in 55 BC, and said: 'The light and heat of the sun; these are composed of minute atoms which, when they are shoved off, lose no time in shooting right across the interspace of air in the direction imparted by the shove'. With hindsight, this seems a remarkably accurate description, for the time; but it does not reflect what most people thought at that time. The idea that sight is associated with something reaching out from the eye persisted for centuries. Plato, who lived from about 428 BC to 347 BC, wrote of a marriage between the inner light and the outer light. Euclid, who lived from about 330 BC to about 260 BC, worried, among other things, about the speed with which sight 'worked'. He pointed out that if you close your eyes, then open them again, even the distant stars reappear immediately in your sight, although the influence of sight has had to travel all the way from your eyes to the stars and back again before you can see them!

Strange though these ideas seem to us, they do not appear to have been seriously challenged, in spite of Lucretius' interest in the work of Epicurus, until the end of the first millennium after Christ. One reason, of course, was the collapse of European civilization into the Dark Ages following the fall of the Roman Empire in the west. The Romans had never been very interested in science anyway, and the scholarship of their times never recovered from the accidental burning of the great library in Alexandria, in Julius Caesar's time, when most of the scientific teaching of the Greeks went up in smoke. More books were destroyed or lost when the Empire collapsed, and such 'scientific' scholarship as remained in Europe consisted, for more than a thousand years, of venerating the ideas of the ancients and trying to preserve what fragments of their teaching remained.

The first scientist to go beyond the ideas of the ancient Greeks in any area of study was an Arab scholar, who lived from about 965 to 1038, at the height of the great Islamic civilization. A great deal of what we know about the ancient world, and its scientific ideas, has come to us from documents which were translated from Greek or other ancient languages into Arabic, and later from the Arabic into European languages. The material reached the Arab world, in many cases, through the Roman Empire in the east – Byzantium, which survived for almost a thousand years after the fall of Rome, until 1453. The relationship between Byzantium and the Arab world was turbulent, to say the least, but it included the exchange of intellectual ideas.

Building on the ideas of the ancients, and improving them (our numbering system, remember, is Arabic), the Arab scholars passed on a rich inheritance to western Europe, an inheritance which played a major part in rekindling the fire of scientific inquiry. The study of light provides a good example of this.

THE FIRST MODERN SCIENTIST

Abu Ali al-Hassan ibn al-Haytham was the greatest scientist of the Middle Ages, and his many achievements were not surpassed for more than 500 years, until the time of Galileo, Kepler and Newton. He became known in Europe (eventually) as Alhazen. He wrote scores of books (what we would now call scientific papers), on a variety of scientific and mathematical topics; but his greatest work was contained in a series of seven books on optics that he wrote either side of the year 1000. This work was translated into Latin (the intellectual and scientific language still used by educated people across Europe until well after Newton's day) at the end of the twelfth century; but it was only published in Europe (still in Latin), as *Opticae thesaurus* (*The Treasury of Optics*), in 1572. It was studied widely, and became a major influence on the thinkers who started the scientific revolution in Europe in the seventeenth century.

Alhazen used several logical arguments to support his contention that sight is not a result of some inner light reaching outward from the eye to probe the world around it, but is solely a result of light entering the eye from the world outside. One of his arguments concerned the familiar phenomenon of after-images. If you stare at a bright light for about half a minute and then close your eyes, you will 'see' an outline of the bright light, usually in a different colour (known as the complementary colour) from the original.[1] Such after-images may even persist, as 'spots before the eyes', after you open your eyes again. Alhazen reasoned that this can only be a result of something from outside affecting the eyes, making such a strong impression that

---

[1] But *never* look directly at the Sun; staring at the Sun for even a short time can result in permanent damage to your eyes.

the effect persists even when the eye is closed, so that light can neither get in nor get out.

There were other examples of what Alhazen saw as the effect of light entering the eye from outside. But the image which was to have the greatest influence on the scientific development of an understanding of the behaviour of light was his discussion of the way images are formed in a 'camera obscura'. The term literally means 'darkened chamber' (or 'dark room'), and the phenomenon was certainly known to the ancients; but the earliest clear description of the phenomenon that we have is in the writings of Alhazen. To see the phenomenon at work, stand in a darkened room on a bright, sunny day, and place a heavy cloth over the window. Make a tiny hole in the cloth – about as big as the ball-point on a ball-point pen – to allow a little light into the room. What you will see is a full-colour image of the world outside, projected, upside-down, onto the wall opposite the curtained window.

The effect is dramatic and entertaining; so much so that even today, in the age of television, some cities (including Edinburgh in Scotland) have modern versions of the camera obscura set up as tourist attractions. The same phenomenon occurs in a pinhole camera, in which the dark 'chamber' may be a shoe box, or something of similar size, with a pinhole in one end and a sheet of tracing paper stuck over the cut-off end of the box opposite the hole, to make a screen. With your head and this screen end of the box kept in the shade (maybe by pulling your coat up over your head) and the pinhole out in the light, you will see an inverted image of the world on the tiny screen. The camera obscura eventually led to (and gave its name to) the photographic camera. But how does it work?

The key point, as Alhazen realized, is that light travels in straight lines. Imagine that there is a tree some distance away, in the garden that the window of a camera obscura faces on to. A straight line from the top of the tree through the hole in the curtain will carry on downwards to a point near the ground on the wall opposite. But a straight line from the base of the tree will go upwards through the hole to strike the wall opposite near the ceiling. Straight lines from every other point on the tree will go through the hole to strike the wall in correspondingly well-determined spots. The result is an upside-down image of the tree (and of everything else in the garden).

Alhazen thought of light as being made up of a stream of tiny particles, produced in the Sun and in flames on Earth, which travelled in straight lines and bounced off objects that they struck. Light from the Sun bounces off the tree in the garden, through the hole in the curtain, off the wall at the back of the darkened chamber and, eventually, into your eyes, which is why you can see the image in a camera obscura. He realized that light cannot travel at an infinite speed, even though it must travel very fast – he thought about the way that a straight stick looks as if it is bent when one end is placed in water, and realized that this effect, refraction, is a result of light travelling at different speeds in water and in air. He also studied lenses and curved mirrors, working out that the curvature of a lens enables it to focus light by refraction.

But Europe wasn't ready for all of this in the eleventh century. The person who first took up the baton passed on by Alhazen was Johannes Kepler, remembered today chiefly for his discovery of the laws that describe the way planets move around the Sun; he lived from 1571 to 1630. Early in the seventeenth century, jumping off from Alhazen's discussion in the *Opticae thesaurus*, Kepler described the human eye in the same terms as a pinhole camera, with light entering through the pupil and forming an image of the outside world on the back of the eye, the retina. For centuries, this left unanswered the puzzle of why we see the world 'right way up' when the image on the retina must be inverted – René Descartes proved that it *is* inverted by taking an eye from a dead ox, scraping the back to make it transparent, and looking at the image formed on the retina. We now know that the human brain automatically corrects for the upside-down image, rather like the way the image on a TV screen can be turned the right way up electronically, even if the TV set itself is upside down.

About this time (Descartes lived from 1596 to 1650), there was an explosion of interest in light. Galileo Galilei, who had been born in 1564 (the same year that William Shakespeare was born) and died in 1642 (the same year that Isaac Newton was born) heard about the invention of the telescope by a Dutch spectacle-maker in 1608, and quickly made his own version, turning it on the heavens and inventing the modern science of astronomy. The microscope was invented soon after the telescope, giving scientists a chance to probe inward, to the world of the very small, as well as outward, into the Universe at large. Using the telescope, Galileo discovered the four largest moons of Jupiter, in 1610; in 1676, studies of the movement of these moons made it possible to measure the speed of light for the first time.

The trick was carried out by the Danish astronomer Ole Rømer, and depended on measuring the times at which the moons were eclipsed by Jupiter itself. The timing of the eclipses seemed to be affected by whether the Earth was on the same side of the Sun as Jupiter was, or on the opposite side; Rømer explained the differences in the eclipse timings as due to the extra time required for light from the moons to reach the Earth when it was on the opposite side of the Sun. Putting some modern numbers in, it takes light more than eight minutes, travelling at 300,000 km per second, to reach us from the Sun, across half the diameter of the Earth's orbit. So the maximum 'delay' in observing an eclipse of one of the moons of Jupiter is twice that – more than a quarter of an hour.

In the same decade that Rømer showed how to measure the speed of light, however, the person who was to transform not just the study of optics but the whole of science first came to the attention of the scientific community in England. In 1672 Isaac Newton published his first scientific paper, and it was about the nature of light.

FROM WOOLSTHORPE TO CAMBRIDGE – AND BACK

Newton very nearly never became a scientist – at least, not a university-educated scientist and Fellow of the Royal Society – at all. He was born prematurely, a tiny, sickly baby, on Christmas Day 1642[1] at Woolsthorpe near Grantham in Lincolnshire; he was not expected to live, and barely survived his first week. His father, also called Isaac, had died before Newton was born – but that may have worked out to Isaac's eventual advantage. When his mother remarried three years after he was born, and went to live in North Witham, the neighbouring village, Isaac was sent to live with his maternal grandparents. No Newton before Isaac had received an education, and it is highly unlikely that he would have been any exception had his father, a yeoman farmer who could not write his own name, lived; young Isaac would have been destined for a farming life himself. But his mother's family, the Ayscoughs, were further up the social ladder than the Newtons. Grandfather James Ayscough was a Gentleman, and Hannah,

---

[1] According to the old-fashioned calendar still in use in England then; on the continent of Europe, following calendar reforms introduced by the Pope to keep the calendar in step with the seasons, it was already 4 January 1643.

Isaac's mother, had a brother, William, who was a graduate of Trinity College, Cambridge, had taken holy orders and had a parish nearby.

Newton had a lonely childhood – his stepfather never took him into his home – but attended local day-schools, receiving the beginning of an education, and saw a way of life distinctly superior to what he might have had as the son of Isaac Newton, farmer. When his stepfather died in 1653, young Isaac's mother returned to Woolsthorpe, and he went back to live with her. His joy at the reunion must have been tempered, though, by the fact that he now had a half-brother and two half-sisters with whom to share her affections. Just two years later, at the age of 12, Isaac was sent off to the grammar school in Grantham, where he stayed in lodgings at the house of the apothecary, a Mr Clark.

The isolation and repeated separation from his mother, on top of the fact that he never knew his father, must have contributed to the unfortunate personality of Newton as a man – secretive, cantankerous, never suffering fools gladly, and often getting involved in enormous academic rows about priority and accusations of plagiarism. But even though he did well at school, and came to be regarded as unusually intelligent (but distinctly odd), Newton still had one more hurdle to surmount before being set firmly on the path to scientific glory. When he was 17, his mother brought him home to Woolsthorpe to learn how to manage the farm, which she intended him to take over from her. He was hopeless at the job. While Hannah despaired of turning her son into a farmer, her brother William urged her to let Isaac go back to school to prepare for entrance to the university. The schoolmaster in Grantham, Mr Stokes, was even more persuasive, offering to let the young man stay in his own house, and to reduce the fees, if he could have his prize pupil back. In 1660, the year Charles II was restored to the throne of England after the 11-year-long parliamentary interregnum, Hannah relented, and Isaac was allowed to return to his studies in Grantham. And he set out for Cambridge, at last, in June 1661. Now, there would be no turning back.

The official curriculum at Cambridge was still, in the 1660s, based on the ancient ideas of the Greek philosophers, notably Aristotle. Newton seems to have followed the prescribed courses reasonably diligently, and he gained his degree in 1665. But at the same time he had been reading the works of more modern thinkers, including Kepler, Galileo and Descartes, educating himself in the new science of the mid seventeenth century. When plague broke out in London in

1665, the University of Cambridge was closed, and Newton went home to Lincolnshire. He stayed there for two years, thinking about what he had learned and developing his own ideas about the way the Universe worked. It was during those two years that he invented the mathematical calculus, developed his theory of gravity, and came up with his theory of light and colour. But none of these was made public for years; it was enough for Newton to solve the problems to his own satisfaction, and it was only with great difficulty that his colleagues later persuaded him (once they had found out what he was up to) to publish the fruits of his labours for all to see.

By the 1660s, there were two rival theories about light. One, espoused by the French physicist Pierre Gassendi (who lived from 1592 to 1655), held that it was a stream of tiny particles, travelling at unimaginably high speed. The other, put forward by Descartes, suggested that instead of anything physically moving from one place to another the Universe was filled with some material (dubbed 'plenum') which pressed against the eyes. This pressure, or 'tendency to motion', was supposed to produce the phenomenon of sight. Some action of a bright object, like the Sun, was supposed to push outwards. This push was transmitted instantaneously, and would be felt by the human eye looking at the bright object.

There were problems with both ideas. If light is a stream of tiny particles, what happens to them when two people stand face to face, looking each other in the eye? And if sight is caused by the pressure of the plenum on the eye, then (as Newton himself pointed out in his notebooks) a person running at night should be able to see, because the runner's motion would make the plenum press against their eyes.

Newton favoured the idea of light being made up of a stream of particles (or corpuscles), not least because of the success of his laws of mechanics in explaining the behaviour of particles. He thought that he could apply essentially the same laws to the motion of planets around the Sun, the flight of a cannonball, or the behaviour of particles of light. He was, in a sense, trying to develop a unified theory of physics, more than 300 years ahead of his time. But in 1661, when Newton went up to Cambridge, developments from Descartes' rival idea were beginning to look more promising.

In its original form, Descartes' theory of light envisaged a steady pressure pushing on the eye. It was only a small step, though, to elaborate this into a theory involving pulses of pressure spreading out from a bright object. These pulses would make waves – not like ripples on a pond, but like the pulses of pressure that would travel through the water of the pond if you slapped the top of the water with your hand (and exactly equivalent to the pressure waves which, we now know, explain how sound spreads outward from its source). In the early 1660s, at least two people, Robert Hooke in England and Christiaan Huygens in Holland, were beginning to think along these lines, moving towards a fully-fledged wave theory of light.

More of Hooke shortly. Huygens deserves something more than just a passing mention, since he ranks second only to Newton among the great physicists of his time – no mean achievement, since Newton is still regarded as the greatest scientist who ever lived.

IN NEWTON'S SHADOW

Huygens was born in The Hague in 1629. His family life was strikingly different from Newton's; his father was a diplomat and poet, coming from a family with a tradition of diplomatic service to the Royal House of Orange. Descartes, who served with the army of the Prince of Orange as a young man, and lived in Holland from 1628 to 1649, was a frequent house guest of the Huygens, and this may have been a factor in Christiaan's choice of career. He was educated in mathematics and law, and groomed to follow the family tradition of a diplomatic career. But in 1649 he returned home, after completing his formal education, and he spent the next 16 years as a gentleman scientist, living off an allowance from his father.

Coming from such a privileged background, and with a family able and willing to indulge his whims, it would have been easy for Huygens to have become a dilettante amateur scientist, dabbling in the study of nature. But he was deeply interested in all facets of science, and made major contributions in several areas. Far from being a dilettante, he was so successful, and so well-known, that when the French Royal Academy of Sciences was founded in 1666 Huygens was invited to work there as one of the seven founding members. He stayed until 1681, when he was forced to return to Holland, partly through ill health and partly through the threat of religious persecution for his Protestant views in Catholic France. He made occasional trips abroad, including a visit to London in 1689, when he met Isaac Newton. He died in The Hague in 1695.

In one respect, Huygens was like Newton. He often delayed publishing his ideas. In his case, though, the main reason was his pernickety obsession with getting everything absolutely perfect, dotting all the *i*s and crossing all the *t*s, before letting his work appear in print. This obsessive attention to detail stood him in good stead when working on his first great contribution to seventeenth-century science, the pendulum clock.

Although Galileo had realized in 1581 that a swinging pendulum kept a regular rhythm, whatever the extent of its swing, nobody had succeeded in using the regularity of the swing of a pendulum to drive an accurate clock until Huygens came up with a practical design in the 1650s. The first clock using his design was built in 1657; a year later, pendulum clocks in church towers were a common sight across Holland. The invention also transformed science by providing an accurate means of keeping time – crucially important, for example, in Rømer's determination of the speed of light, and in other areas of astronomy. Going one step better than the tower clock, in 1674 Huygens also developed the first working watch, driven by a spring and controlled by a balance-wheel instead of a pendulum (although Hooke had independently come up with the same idea, Huygens had the first working model).

Huygens also designed telescopes, and made astronomical observations. He discovered Titan, the largest moon of Saturn, in 1655, and was the first person to describe the nature of the rings of Saturn correctly. Through his astronomical work, and his efforts to build better telescopes, Huygens became interested in the nature of light, and this led to his greatest achievement, a fully worked-out wave theory of light, which was essentially complete by 1678 but not fully published until 1690. His theory was able to explain the way light reflects from a mirror, and the way it is refracted (bent) when it passes from air into glass or water. Building from Descartes' idea, he envisaged light as a kind of jostling motion of particles, nudging each other and spreading a disturbance out from its source as a spherical pressure wave. And his theory made one particularly significant prediction: that, in order to explain refraction, light should travel more slowly in a denser medium (such as glass or water) than it does in a less dense medium (such as air).

Huygens's misfortune, though, was to have his reputation dimmed by the shadow of Newton. Newton's breathtaking achievements in 'natural philosophy' – his laws of motion and theory of gravity – were published in 1687, in the famous *Principia*. Some of his ideas about light had appeared in print 15 years earlier, although his complete theory was not published until 1704, for reasons that will shortly become clear. Largely on the strength of Newton's justified reputation as the greatest scientific genius, in the eighteenth century his ideas about light, as well as his laws of motion and theory of gravity, were widely regarded as gospel. One aspect of Newton's views about light was that it came in the form of particles – so, obviously, Huygens's theory must be wrong. But even geniuses sometimes make mistakes, and in fact the particle hypothesis was by no means the most important aspect of Newton's theory of light. It was his theory of colours that first drew him to the attention of the scientific world of the time – and like Huygens's work on light, Newton's theory had astronomical connections.

NEWTON'S VIEW OF THE WORLD

The important thing about Newton's theory of colours is not just that he was right, but the way in which he arrived at his conclusions. Before Newton, the way philosophers developed their ideas about the natural world was largely through pure thought. Descartes, for example, thought about the way in which light might be transmitted from a bright object to the eye, but he did not carry out experiments to test his ideas. Of course, Newton was not the first experimenter – Galileo, in particular, pointed the way with his studies of the way in which balls rolled down inclined planes, and with his work on pendulums. But Newton was the first person to express clearly the basis of what became the scientific method – the combination of ideas (hypotheses), observation and experiment on which modern science rests.

Newton's theory of colours emerged from experiments he carried out during his enforced sabbatical from Cambridge. By 1665, the fact that a ray of sunlight could be turned into a rainbow-like spectrum of colours by passing it through a triangular glass prism was well known. The standard explanation of the effect was based on the Aristotelian idea that white light represented a pure, unadulterated form, and that it became corrupted by passing through the glass. When the light enters the prism, it is bent, and then follows a straight line to the other side of the triangle, where it bends again as it emerges into the air. At the same time, the light is spread out, from a single spot of white light into a bar of colours. Working downwards from the point of the triangle, the light at the top is bent least, and travels the shortest distance through the glass, emerging as red. Lower down, where the triangular wedge of glass is wider, light which has been bent slightly more as it enters the prism travels further through the glass, and emerges into the air on the other side as violet. In between, there are all the colours of the rainbow – red, orange, yellow, green, blue, indigo and violet. Using a prism held up to the ray of light entering a darkened room through a small hole in the curtain (rather like the camera obscura set-up), the spectrum of colours can be displayed on the wall opposite the window.

On the Aristotelian view, white light that had travelled the shortest distance through the glass was modified least, and became red light. White light that had travelled a little further through the glass was modified more, and became yellow – and so on all the way down to violet.

Newton actually tested these ideas, using both prisms and lenses which he ground himself, trying to minimize the colour change by making lenses in different shapes. He was the first person to distinguish the rays of different colours, and he named the seven colours of the spectrum (he deliberately chose to divide the spectrum into seven colours because seven is a prime number with alleged mystic connotations; if you find it hard to distinguish a separate colour 'indigo' between blue and violet in the rainbow, you are by no means alone!).

But the most important experiment Newton carried out at this time simply consisted of placing a second triangular wedge of glass behind the first prism but the other way up. The first prism, point uppermost, spread a spot of white light into a rainbow spectrum. The second prism, point downwards, combined the spread-out colours of the spectrum back into a spot of white light. Even though the light had passed through a further thickness of glass, it had not become more corrupted, but had returned to its former purity.

As Newton realized, this shows that white light is not 'pure' at all, but is a mixture of all the colours of the rainbow. Different colours of light bend by different amounts when they are refracted, but all the colours are present in the original white spot of light. It was a revolutionary idea, both because it overturned a basic tenet of Aristotelian philosophy and because it rested upon the secure foundation of experiment. But Newton did not hurry to tell the world of his discovery. Instead, the insight into the nature of light that he gained in 1665 led him to try a new approach to telescope-making.

One of the problems associated with telescopes that use large lenses (refracting telescopes) is that the lenses also tend to split up white light into the spectrum of colours. This produces coloured fringes and blurs the image of the object being looked at, a phenomenon known as chromatic aberration, which is particularly inconvenient if you want to study the stars. Newton realized that it would be difficult to construct a lens system which did not suffer from this problem (difficult, but not impossible; so-called 'achromatic' lenses, made of two or more pieces of glass with different refracting properties, can be used to make telescopes which do not produce chromatic aberration). So he designed and built a telescope using a curved mirror instead of large lenses – a reflecting telescope.

Newton's reflector used the neat idea of bouncing light from a large curved mirror at the back of the telescope up to a small flat mirror, tilted at 45 degrees, which diverted the image out through a hole in the side of the telescope tube. The observer could study the stars by looking in through this hole, without their head getting in the way of the light from the stars. The idea was brilliant in its simplicity, but making an accurate mirror with the available material was a laborious task which Newton, an expert model-maker, carried out himself. The result was an instrument about 20 cm (6 in) long, which produced images nine times bigger than those of a refractor four times longer – and without chromatic aberration.

By now, with the threat of plague receding, the University had reopened, and Newton was back in Cambridge. He had been elected a Fellow of Trinity College in 1667. In the same year, England was at war with Holland, and the Dutch fleet successfully attacked the English in the Thames. The sound of gunfire was heard in Cambridge, and everyone knew its cause; Newton impressed his colleagues by asserting (correctly, as it turned out) that the Dutch had won the battle. His reasoning was that the sound of guns had got louder and louder, implying that the battle was coming closer and closer, and therefore that the English were in retreat.

By 1669, Newton was becoming known outside Cambridge for his work in mathematics, and in that year the first Lucasian Professor of

Mathematics, Isaac Barrow (who had been appointed in 1663) resigned, effectively in Newton's favour. Barrow, although an accomplished mathematician, had ambitions elsewhere, and soon became first chaplain to the King and then Master of Trinity College. His influence over the executors of the will of Henry Lucas, which had established the Lucasian chair, was amply sufficient to ensure that his successor in that post was another Trinity man, who had just made his mark as a mathematician of note.

The appointment secured Newton's position in Cambridge, but required him to deliver a regular series of lectures. The topic he chose for his first course of lectures was not mathematics but optics and the theory of colours, with special reference to the problem of chromatic aberration in telescopes. At the same time, he was proudly showing off his new telescope to colleagues in and around Cambridge; indeed, the earliest surviving copy of a letter by Newton is one written (to an unknown correspondent) in February 1669, and is largely a description of the telescope.

News of the remarkable instrument reached the Royal Society (which had been formally established in 1662, although it had existed on an informal basis since 1645) late in 1671, and the Secretary of the Society, Henry Oldenburg, asked to see the telescope. Barrow delivered it to the Society, in London, for Newton, and in January 1672 Oldenburg wrote a fulsome letter to Newton, expressing their appreciation of his invention and informing him that news of the telescope had been passed on to Huygens, at that time in Paris. So Newton's fame spread, as the inventor of a new kind of telescope, to the mainland of Europe; on 11 January 1672 he was elected a Fellow of the Royal Society on the strength of this invention, and a few weeks later his first physics paper was published, in the form of a letter to Oldenburg setting out his theory of colours. It appeared in the *Philosophical Transactions* of the Society on 19 February 1672, and led to the first of Newton's famous academic rows.

Robert Hooke, who had been born in 1635 and was to live until 1703, was at the time Curator of Experiments at the Royal Society. He was an established figure in the scientific world, had his own ideas about light and colour (his own wave theory of light, less complete than Huygens's, was published in 1665), and was always eager to claim priority for any of his work. He responded to Newton's letter in patronizing terms, dismissing the notion that light could be made of corpuscles and failing to appreciate that the theory of colours did not, in fact, depend upon the corpuscular hypothesis. In acrimonious exchanges, Hooke implied that what was original in Newton's theory was wrong, and that what was right in Newton's theory was not original.

The resulting row had two effects. First, it led to Newton retreating from the world of science at large and keeping himself to himself in Cambridge, refusing to publish anything much at all for years (and hugging his complete theory of optics to himself until Hooke died, when he could publish it in full and be sure of having the last word). Secondly, it led to Newton's famous remark 'If I have seen further it is by standing on the shoulders of Giants' – a cutting reference to the fact that Hooke was small of stature, and implying that he was small in intellect as well.[1]

In correspondence with another critic of his theory of colour, however, Newton provided an insight into his method of working – what became the scientific method. The French Jesuit Ignace Gaston Pardies wrote to Newton from Paris, making several points about the theory in what Newton evidently regarded as a suitably respectful way. Instead of dismissing Pardies as a fool, he wrote back explaining his arguments more fully, and saying:

> The best and safest method of philosophizing seems to be, first to enquire diligently into the properties of things, and to establish those properties by experiments and then to proceed more slowly to hypotheses for the explanation of them. For hypotheses should be employed only in explaining the properties of things, but not assumed in determining them; unless so far as they may furnish experiments.[2]

This is what science is all about. No matter how wonderful your theory may be, if it does not match up with the results of experiments it cannot be correct. For example, Newton's theory of light (perhaps we should say 'hypothesis', although Newton bridled when Hooke used the term about these ideas) also explains refraction as due to a change in the speed of light when it moves from one medium to another. But unlike Huygens's theory, the corpuscular theory requires light to travel *faster* in a more dense medium. Here is a clear-cut way to distinguish between the two ideas; had Newton lived to see the

[1] The Giants quote has nothing to do with Newton's theory of gravity, but comes from a letter to Hooke written in 1675, 12 years before the *Principia* was published. For details of the row with Hooke, see Gribbin, *In Search of the Edge of Time*, Chapter One.
[2] Quoted in Westfall, *Never at Rest*, p. 242.

experiments carried out which show that light does, in fact, move more slowly in a more dense medium, he would surely have accepted the evidence that light travels as a wave.

As well as establishing the scientific method of investigating the way the world works, Newton (along with Huygens and their contemporaries) established the first scientific paradigm, or model, of reality. This showed that the Universe obeys precise rules, or laws, and that events as different as the motion of the planets around the Sun and the bending of a light beam can be explained by the application of these rules, rather than by the whims of capricious gods.

The image handed down to us by the giants of seventeenth-century science is often referred to, rightly, as that of a 'clockwork Universe', obeying inexorable laws. But the correct image is not that of a modern clock or wristwatch, ticking away the seconds one by one. Rather, we should imagine a great cathedral clock of the seventeenth century, driven by a huge pendulum in accordance with Huygens's design, with many interconnecting cogs and gearwheels that do not just tick away the time but which drive a complicated mechanism to set in motion sophisticated tableaux involving moving figures of the saints, striking bells and other mechanical activity at appointed hours. *That* is the kind of complex clockwork that seventeenth-century science envisaged underpinning the dance of the planets around the Sun and other natural phenomena.

Newton's legacy combines the idea that the behaviour of everything in the Universe is predictable, in the same way that the actions of the figures in the clockwork tableaux of cathedral clocks are predictable, *and* the fact that relatively simple laws intelligible to human brains are all that is required to understand what makes the Universe tick. In the light of these achievements, the fact that the next step forward in understanding the nature of light seemed to show that Newton's corpuscular theory was wrong pales into insignificance. But it was still an important step.

YOUNG IDEAS

Direct evidence that light travelled as a wave already existed in Newton's day. But the evidence was slight, the work was little known, and the explanation of the phenomenon was incomplete. It came from the work of the Italian physicist Francesco Grimaldi (1618–63), who, like Newton, studied the behaviour of a beam of sunlight let into a darkened room through a small hole. He found that when the beam went through a second small hole and onto a screen, the spot of light on the screen was slightly larger than the hole, and had coloured fringes. The light had spread out slightly, with different colours spreading by different amounts, as it passed through the small hole.

He also found that if a small obstruction was placed in the beam of light, the resulting shadow on the screen had coloured edges, where light had leaked into the fringes of the shadow. It had spread *in* to the shadow, and again different colours had spread in by different amounts. Both effects are very small, but can be clearly detected by careful observation and measurements. Grimaldi gave this family of phenomena the name diffraction – a third way, alongside reflection and refraction, that light could be bent. But Grimaldi's work on diffraction was only published in 1665, two years after his death, and he was not around to argue the case for the wave theory by the time Newton's ideas had seized the imagination of the scientific community. Hooke also found that light does not travel in perfectly straight lines, but leaks into the edges of the shadow of an object placed in its path. But we have seen why *he* was not around to argue the case for the wave theory when Newton published his complete theory of optics.

Although Newton's ideas dominated scientific thinking in the eighteenth century, the wave theory of light was not without some supporters. The most notable proponent of the idea was the Swiss mathematician Leonhard Euler, who was born in Basle in 1707, and was just a few weeks short of his twentieth birthday when Newton died in 1727. Euler was one of the great mathematicians of all time, interested both in pure mathematics and practical applications in the study of tides, fluid mechanics, predicting the movement of heavenly bodies using Newton's laws, and in other areas. Even a great scientist, though, can sometimes make a foolish mistake. While he was Professor of Mathematics in St Petersburg, in Russia, in the 1730s, Euler lost the sight of his right eye through looking at the Sun during his astronomical work. Thirty years later, when he was back in St Petersburg as Director of the Academy of Sciences (this was at the time of Catherine the Great), he went blind in his other eye as a result of a cataract; but he stayed in the post, carrying out all his duties and responsibilities, until he died in 1783. He remained active as a mathematician during the last 15 years of his life, carrying out calculations entirely in his head, and dictating his discoveries to an assistant. On the day he died, at the age of 76, he had spent part of his time calculating the laws of ascent of hot-air balloons, which had recently been invented.

Euler's theory of light was published in 1746, in between his two spells in St Petersburg, when he was working at Frederick the Great's Academy of Sciences in Berlin. He pointed out all the difficulties with the idea that light travelled as a stream of tiny particles (including the difficulty of explaining diffraction in this way), and made the specific analogy between the vibrations of light and the vibrations of sound waves. By now, the term used for the medium doing the vibrating had been changed from 'plenum' to 'ether'. In a letter written in the 1760s, Euler said that sunlight is 'with respect to the ether, what sound is with respect to air', and described the Sun as 'a bell ringing out light'.[1] But still the world was not convinced. Significantly, the wave theory only supplanted the corpuscular theory when new experiments were carried out to test the idea. Newton's corpuscular theory of light would be overturned as a direct result of the application of the Newtonian paradigm of the way science ought to be done.

The first step was taken by the British physicist Thomas Young, who was born in 1773 and 10 years old when Euler died. The age might seem insignificant; but Young was a genuine infant prodigy, who had crammed more into the first ten years of his life than many people achieve in a lifetime. He could read English at the age of two, and devoured books supplied by a doting grandfather. At six, he moved on to Latin, and then other languages; by the age of 16 he understood Latin, Greek, French, Italian, Hebrew, Chaldean, Syriac, Samarian, Arabic, Persian, Turkish and Ethiopic. As the list of his languages suggests, Young was interested in archaeology and ancient history from an early age; in fact, he was interested in just about everything. In 1792, at the age of 19, he began to study medicine, intending to join a wealthy great-uncle in his London practice. He studied in London, Edinburgh and Göttingen, receiving his MD from the German university in 1796. But in his first year as a medical student Young had explained the way in which the focusing mechanism of the eye works, with muscles changing the shape of the lens of the

[1] Quoted in Zajonc, *Catching the Light*, p. 99.

eye, and as a result of this work he was elected a Fellow of the Royal Society at the age of 21, while still a student.

After receiving his MD, Young travelled in Germany, then worked in Cambridge for two years, carrying out a variety of scientific studies and gaining the nickname 'Phenomenon Young' for his versatility. In 1800 he returned to London and did set up a medical practice, eventually becoming a physician at St George's Hospital, a post he held from 1811 until his death in 1829. But medicine remained just one of many interests.

Young explained astigmatism as caused by irregularities in the curvature of the cornea of the eye, was the first person to realize that colour vision is produced from a combination of just three colours (red, green and blue) influencing different receptors in the eye, carried out important work in physics (including the first estimate of the sizes of molecules), and served as Foreign Secretary to the Royal Society (no doubt at least some of his languages came in handy). From 1815 onwards, he returned to his early interest in ancient history, publishing papers on Egyptology and helping to decipher the Rosetta Stone, which had been found near the mouth of the Nile in 1799 (Young was probably the leading light in deciphering the ancient scripts on the Rosetta Stone, but did not receive full credit at the time because his main work on the problem was published anonymously, as a supplement to the *Encyclopaedia Britannica* for 1819). But in spite of all this, and more, Young is best remembered for his investigation of the phenomenon of interference of light.

His first experiments in interference were carried out during his time in Cambridge, from 1797 to 1799; they continued when he returned to London, and in the early years of the nineteenth century Young presented clear and accurate accounts of these experiments, arguing in favour of the wave theory of light, to a sceptical British scientific community. Young carried out (indeed, invented) the basic interference experiment, described in the Prologue to this book, both with two round pinholes and with two narrow slits. Even better, in some ways, he explained some of Newton's own experiments with light in terms of the wave theory. He realized that each colour of light corresponds to a particular wavelength, and that the amount by which light bends when it is refracted or diffracted depends on its wavelength. Armed with this information, he used Newton's own data to calculate that the wavelength of red light is 6.5×10^{-7} metres, and that of violet light is 4.4×10^{-7} metres. Both figures are in good agreement with the values accepted today, which shows both how good a theorist Young was and how accurate an experimenter Newton had been. The figures also show why it took so long to prove that light does travel as a wave. These wavelengths are tiny – about half a millionth of a metre long – and the size of the diffraction effect is, in round terms, comparable to the wavelengths involved. The amount by which light bends when going past the edge of an object is only a few millionths of a metre. But only waves, albeit tiny waves, can explain what happens when light passes through the double slit experiment.

In 1807, just 80 years after Newton had died, Young summed up the experiment with two holes in the following words:

> The middle [of the pattern] is always light, and the bright stripes on each side are at such distances, that the light coming to them from one of the apertures must have passed through a longer space than that which comes from the other by an interval which is equal to the breadth of one, two, three, or more of the supposed undulations [wavelengths], while the intervening dark spaces correspond to a difference of half a supposed undulation, of one and a half, of two and a half, or more.[1]

The description is exactly right. Ten years later, Young suggested that light waves are transverse (vibrating from side to side across the direction of motion), rather than longitudinal (pushing and pulling in the direction of travel, like sound waves, or the 'waves' you make in its bellows while playing an accordion).

You might think that all this would have been convincing proof of the wave nature of light. But even Young was unable to persuade the scientific establishment of his day that Newton had been wrong about the nature of light. As well as the widespread feeling that there was something distinctly unpatriotic, perhaps even dishonourable, in the suggestion that Newton had been wrong about anything, the idea that you could make darkness by *adding* two rays of light together was almost incomprehensible to many of Young's colleagues. We are used to the idea, and the experiment with two holes, described purely in terms of waves, seems to make perfect common sense to us. But in the early nineteenth century common sense said that adding two rays of light together must always increase their brightness; the idea that adding two rays of light could produce darkness was, in the words of one of Young's contemporaries, 'one of the most incomprehensible

[1] Quoted in Baierlein, *Newton to Einstein*, p. 95.

suppositions that we remember to have met with in the history of human hypotheses'.[1] Perhaps appropriately, the final 'un-British' overthrow of Newton's corpuscular hypothesis was carried out by a Frenchman, ignorant of Young's work (which is not too surprising, remembering that Britain and France were at war, except for one brief interruption, from 1799 to 1815).

Augustin Fresnel was born in 1788 in Broglie, Normandy. He qualified as an engineer in 1809, and worked for the government on road projects in various parts of France. His interest in optics was only a hobby, and he was not part of the circle of scientists who might have learned of Young's work, even during the time of war between Britain and France. When Napoleon was defeated and exiled to Elba, Fresnel 'came out' as a Royalist – and when Napoleon returned from Elba for the Hundred Days, in 1815, Fresnel either quit his position in protest, or was sacked (the reports are not clear). Either way, he was placed under house arrest at his home in Normandy, where he developed his half-formed ideas about light into a fully-fledged theory. When Napoleon was finally overthrown, Fresnel went back to his engineering work, and optics became, once more, a part-time interest.[2] But in his spare time, and that one period of enforced leisure, he did enough to ensure that the particle theory of the propagation of light was finally laid to rest.

FRESNEL, POISSON, AND THE SPOT

Although it is not too surprising that Fresnel did not, in 1815, know of Young's work, it is slightly more surprising that he didn't know about the work of Huygens or Euler. Nevertheless, that seems to have been the case. His wave theory was entirely his own work, and was developed as the simplest explanation for the phenomenon of diffraction. The clinching piece of evidence came from an experiment that is in some ways even simpler than the experiment with two holes, but which is even more surprising.

[1] Henry, Lord Brougham, quoted in Zajonc, *Catching the Light*, p. 110.
[2] Some of this work had practical applications; in 1820 Fresnel developed a type of lens, made of a series of concentric rings. This type of lens, still named after him, is used in concentrating the beam of light from a lighthouse and in other applications.

You can, in fact, see the stripey pattern produced by diffraction and interference using just a single slit (one hole) – and you don't need any sophisticated scientific equipment to do the trick. Try holding your hand up in front of your face, with the fingers very nearly touching. Look through the narrow gap between the middle two fingers at a bright light, and gently squeeze the fingers together so that the gap between them gets narrower. Just before the gap disappears entirely, you will see a pattern of dark stripes in the 'slit' between your fingers. You may see just one or two dark lines in the middle of the gap, or, if you are careful, several stripes.

Physicists can do the same thing using a single narrow slit and casting the light from the slit onto a screen. The full explanation of the phenomenon is straightforward, but involves a modest amount of calculation; you can think of it as being caused by the light bending round either edge of the single slit and travelling by two different paths, each with a different number of wavelengths in, to your eyes, or to the screen behind the slit. Fresnel's key proof of the wave nature of light turned this single-slit diffraction experiment inside out, using a single small obstruction in a beam of light, and looking at interference effects occurring in its shadow, as a result of light bending around the edges of the object. This is a bit like the way waves wash around a rock to disturb the water behind the rock, but on a much smaller scale.

In 1817, with the Napoleonic wars at last at an end, the French Academy of Sciences, stimulated by the work of Young (but possibly still ignorant of the work of Fresnel), decided to attempt to resolve the question of the nature of light once and for all. They offered a prize for anyone who could provide the best experimental study of the phenomenon of diffraction, *and* supply a theory that satisfactorily explained the behaviour of light in the experiment. Although the competition was open to anybody, not just the French, it attracted only two entries. One was, apparently, a crackpot contribution; history does not tell us even the name of the person who submitted the work, let alone the details of the experiment he suggested. The other entry came from Fresnel, in the form of a comprehensive paper running to 135 pages. Of course, he won – but not without opposition from the judges, who met in March 1819 to announce their decision. They included the mathematician Siméon Poisson, the physicist Jean Biot and the astronomer Pierre Laplace, all strong supporters of the Newtonian theory.

Fresnel was no mean mathematician, and used the techniques of calculus developed by Newton and by Wilhelm Leibniz to set up a formal description, in mathematical language, of the behaviour of light under various diffraction situations. But the equations were sometimes so complicated that even Fresnel could not solve them completely to work out the exact details of the way light would bend in a particular set of circumstances. Poisson, however, as well as being a confirmed Newtonian, was a red-hot mathematician. He lived from 1781 to 1840, and made major contributions to probability studies, calculus, the behaviour of electricity and magnetism, and many other areas of research. He seized upon one of Fresnel's examples, solved the relevant equations, and presented his fellow judges with what seemed to be a *reductio ad absurdum* that would pull the rug from under the wave theory once and for all.

The notion that coloured fringes on the *edge* of a shadow might be produced by the diffraction of light waves did at least match up with common-sense ideas about the way waves behave. But Fresnel's theory, Poisson's calculations showed, said that there would be a tiny bright spot *directly behind* a small round object placed in a beam of light, in the exact centre of the shadow of the object. Light waves bending round the edges of the obstruction ought to combine to produce this single spot of brightness in the centre of the dark shadow. Absurd! As Poisson himself described the results of his calculations:

> Let parallel light impinge on an opaque disk, the surroundings being perfectly transparent. The disk casts a shadow – of course – but the very centre of the shadow will be bright. Succinctly, there is no darkness anywhere along the central perpendicular behind an opaque disk (except immediately behind the disk). Indeed, the intensity grows continuously from zero right behind the thin disk. At a distance behind the disk equal to the disk's diameter, the intensity is already 80 per cent of what the intensity would be if the disk were absent. Thereafter, the intensity grows more slowly, approaching 100 per cent of what it would be if the disk were not present.[1]

But, of course, as good Newtonians the judges had no intention of using mere logic and common-sense reasoning to condemn Fresnel's theory. Using the by now standard Newtonian method of testing hypotheses by experiment, the chairman of the judges, François Arago,

---

[1] Quoted in Baierlein, *Newton to Einstein*, p. 102.

arranged for the experiment to be carried out. He found that there is, indeed, a tiny bright spot of light (known to this day as the Poisson spot) exactly in the middle of the shadow. It shows up for small balls, such as BB shot, as well as for circular disks. Fresnel was right; Newton was wrong. So Arago reported to the Council of the Academy of Sciences at that March 1819 meeting that:

> One of your commissioners, M. Poisson, had deduced from the integrals reported by the author [Fresnel] the singular result that the centre of the shadow of an opaque circular screen must, when the rays penetrate there at incidences which are only a little more oblique, be just as illuminated as if the screen did not exist. *The consequence has been submitted to the test of a direct experiment, and observation has perfectly confirmed the calculation.*[1]

This is the heart of the matter. Theories are only valid if they are borne out by experiment, and whatever the relevant experimental results tell us must be 'true' and be incorporated into any good theory. No matter how bizarre those experimental results may be – like the dual nature of electrons discussed in the prologue – they *cannot* be swept under the carpet and left out of our theories.

Of course, with the endorsement of the prize committee Fresnel's fame was assured. He worked with Arago on some aspects of the transverse wave theory which explained long-standing puzzles about the polarization of light, a significant step forward in establishing that light waves are indeed transverse. He also suggested an experiment to measure the speed of light in water; the experiment was carried out in 1850, and showed that, just as the wave theory had predicted, light moved more slowly in water than in air – but by then nobody needed convincing that light travelled as a wave. Fresnel was elected to the French Academy of Sciences in 1823, and made a Fellow of the Royal Society in 1827 just before he died of tuberculosis, a hundred years after the death of Newton. Fresnel was only 39 years old when he died, and was outlived by Young, who died in 1829 a month short of his 56th birthday. Two years later, the man who was finally to explain how light waves work was born in Edinburgh, in Scotland. But James Clerk Maxwell's explanation of the nature of light built from a theory of the way electricity and magnetism interact with one another that was already being developed in the 1820s, while both Young and Fresnel were still alive.

---

[1] Baierlein, *Newton to Einstein*, p. 103; the italics are mine.

THE BOOKBINDER'S APPRENTICE

Michael Faraday, who was born in 1791, became the greatest experimental scientist of the nineteenth century, overcoming a humble
family background and a lack of formal education by persistence,
ability, and just a little bit of luck. He was the third of four children,
the son of a poor blacksmith at Newington, in Surrey. At the time,
this was still in the country, although it has now been swallowed up
by London, as part of the borough of Southwark. The family later
moved to north London, and at the age of 13 Faraday became an
errand boy working for a bookseller and bookbinder. His rudimentary
education had at least taught him to read, although he was almost
entirely ignorant of mathematics, and, surrounded by books, he began
to devour them voraciously. His employer, a French immigrant who
had fled the revolutionary turmoil across the English Channel, encouraged him, and took him on as an apprentice bookbinder. Faraday
learned his trade well over the next seven years, developing manual
skills which were to prove useful in his later career as an experimental
scientist; he also read everything he could lay his hands on, and was
particularly fascinated by the article on electricity in the *Encyclopaedia
Britannica*.

In 1810, at the age of 19, Faraday joined the City Philosophical
Society, attending regular lectures on scientific topics. He learned the
basics of physics and chemistry, and took part in experimental work
himself. He also took detailed, accurate and neat notes of the lectures,
which he bound up himself into book form. The books became his
passport to a career in science.

Faraday's employer, M. Ribeau, used to show the volumes of
Faraday's lecture notes to his customers in the shop, and one of them,
impressed by Faraday's enthusiasm for science, arranged for the
bookbinder's apprentice to attend lectures being given by Sir Humphry
Davy at the Royal Institution. Davy was a superb lecturer, and the
most famous scientist in England at the time. He had been born in
1778, and among other things developed the use of nitrous oxide
('laughing gas') as a medical anaesthetic. His greatest practical achievement was the invention of a safety lamp that miminized the risk of
igniting the natural gas often found in deep coal seams; the Davy lamp
became the standard design for use in coalmines.

Faraday, already deeply fascinated by science, was fired into even greater enthusiasm by Davy's lectures. He was just coming to the end of his apprenticeship (in 1812), and decided that he wanted a career in science, not as a bookbinder. He wrote up Davy's lectures and bound them into book form, and sought any kind of scientific employment that might be available, totally unsuccessfully. No prospective employer took the unemployed bookbinder seriously as a prospective scientist. Even the one contact Faraday was able to make did not, at first, look like leading to anything permanent. When Davy was temporarily blinded as a result of an explosion in his laboratory, Faraday was able to act as his helper for a few days, and subsequently sent him the bound notes of Davy's own lectures, with a letter begging for permanent employment. Although Davy was flattered, it was to no avail – there simply were not any jobs available at the Royal Institution.

But then came the stroke of luck. Davy's assistant got into a fight, and was sacked for brawling. The job was offered to Faraday, who started work at the Royal Institution on 1 March 1813, at the age of 21. In many ways, Davy was not an ideal employer. He was snobbish, jealous and dismissive of the work of other scientists, and with an excitable temperament. The job itself included, for the first three years, acting as Davy's valet on long trips through Europe. But although the post was menial and the financial rewards minimal (a guinea a week – less than he had been earning as a bookbinder – and two rooms in the attic at the Royal Institution), Faraday was meeting great scientists, and watching one of them at work. He began (in 1816) to publish scientific papers, was the first person (in 1823) to liquefy several gases (including chlorine), was elected a Fellow of the Royal Society in 1824 (in spite of opposition from Davy, who was then President of the Society), and discovered benzene, isolating it from natural oil in 1825. In the same year, he was appointed director of the laboratory at the Royal Institution, and a year later he began to give regular Friday evening lectures that became a Victorian institution in their own right. His fame and success, in fact, began to outstrip those of Davy, who became bitter at the way his protégé had eclipsed his own work; but Davy died young, in 1829, and from then until 1861, when he retired to live in a house at Hampton Court which had been provided for him by Prince Albert in 1858, Faraday and the Royal Institution were almost synonymous. He died in 1867, a month before his 77th birthday. He had the surely unique record of having turned down not only the offer of a knighthood, but also (twice!) the presidency of the Royal Society. 'I have always felt,' he said, 'that there is something degrading in offering rewards for intellectual exertion, and that societies or academies, or even kings and emperors, should mingle in the matter does not remove the degradation.'

In all of his long career, and in spite of many other achievements, Faraday's greatest contribution to science was the insight he provided into the nature of electricity and magnetism. This not only paved the way for an understanding of how light moves; it also provided physics with a paradigm that remains at the heart of the modern understanding of the way the Universe works, the concept of the field of force.

FARADAY'S FIELDS

Faraday's first important investigations into electricity and magnetism were carried out as early as 1821. In the previous year, Hans Oersted, in Copenhagen, had reported the surprising discovery that electricity flowing through a wire would cause a deflection in the magnetic 'needle' of a small compass placed nearby. Clearly, the flow of electricity along the wire was exerting a magnetic influence. André Ampère, whose name is immortalized in the unit we use to measure electric current, showed that two parallel wires attract each other if they both carry an electric current flowing in the same direction, but repel each other if the currents are flowing in opposite directions. And François Arago, whom we have already met, found that a spinning copper disk would make a compass needle placed just over the disk deflect.

The editor of the *Philosophical Magazine* asked Faraday to look into these mysteries and explain them to his readers. Faraday carried out experiments, and came up with the idea that circular 'lines of magnetic force' are wrapped around a wire carrying an electric current. He devised and built a system in which a suspended wire carrying an electric current moved in a circle around a fixed magnet, and in which a suspended magnet moved in a circle around a fixed wire carrying a current, each pushed by the magnetic field. This is the basic principle behind the electric motor and (eventually) the dynamo, or electric generator. If electricity could generate magnetism, Faraday reasoned, then magnetism ought to be able to generate electricity.

He proved this in 1831; the most straightforward example of his 'electromagnetic induction' occurs when an ordinary bar magnet is pushed into or pulled out of a coil of wire. As long as the magnet is moving, an electric current flows in the wire. Faraday had proved both that moving electricity creates magnetism, and that moving magnetism creates electricity. The word *moving* is crucial here; the delay in finding the second effect had been because at first Faraday had expected to find that a *steady* magnetic field would induce an electric current in a nearby wire.[1]

Now he could explain the mystery of Arago's disk – the motion of the copper conductor through the magnet's influence set up an induced electric current in the disk, and the induced current in turn created a magnetic influence which deflected the magnet, in an early example of the process known as feedback. Using a variation on the theme in which a copper disk rotated between the poles of a large magnet, and two wires brushed the surface of the disk, one at the centre and one at the rim, Faraday built the first electricity generator in October 1831.

Faraday thought long and hard about how to explain these phenomena. He was still no mathematician, but he had an inspired ability to think pictorially. He came up with the revolutionary idea that instead of a material ether, or plenum, filling the Universe and transmitting influences mechanically, through the physical interaction of tiny objects jostling against one another, the electric and magnetic forces, and even gravity, could be described in terms of 'lines of force', reaching out through empty space and interacting with one another. Instead of regarding atoms as tiny lumps of solid, impenetrable matter, he said, we should regard them as the centres of concentrations of forces – no more, and no less.

The concept of lines of force is familiar to anyone who has carried out, or watched, the schoolroom experiment of sprinkling iron filings onto a sheet of paper and holding a bar magnet under the paper. The filings do, indeed, arrange themselves in lines, curving between the two poles of the magnet. But this really was a spectacularly new idea in Victorian Britain, especially when applied to *all* the known forces of nature, and Faraday thought long and hard before airing it in two lectures at the Royal Institution, the first in 1844 and the second in 1846. The second lecture, apparently, was an off-the-cuff affair – Charles Wheatstone had been due to give a lecture that day, but had

[1] It indicates how successfully Faraday's ideas have become a part of modern science that it is almost impossible to describe this work concisely without using terms like 'field', which were only introduced later. Today, 'everybody knows' about fields, and lines of force are common sense – just as the ether was common sense to earlier generations!

an attack of stage fright and ran off at the last minute, leaving Faraday with no choice but to fill the gap. After summarizing the talk he had expected Wheatstone to give, and with time on his hands, Faraday extemporized about his ideas concerning lines of force.

In a classic example of a 'thought experiment', Faraday asked people to imagine the Sun sitting alone in space. What would happen if the Earth were suddenly, by magic, to be placed at its appropriate distance from the Sun? How would it 'know' that the Sun was there? He argued that, even before the Earth was put into place, the Sun's influence would extend through space, in the form of gravitational lines of force. The response of the Earth to the Sun's gravitational field is a response to the existence of the lines of force *at the locality of the Earth*, he said, and not to the remote presence of the Sun itself. As far as the Earth is concerned, the lines of force (the field) are the reality. In the same way, Faraday argued, both magnetic and electric lines of force thread their way through the Universe. These fields, as they are now called, were the reality, and matter itself – atoms – was merely associated with places where the fields were concentrated.

In his 1846 lecture, Faraday went further. He argued that light could be explained in terms of vibrations of the electric lines of force. After all, by then it was well known that light was a form of wave – a vibration. Faraday lacked the mathematical skills to develop this idea into a full explanation of the way light moves, but he had a clear physical picture of what might be going on, as he spelled out when the ideas presented in the 1846 lecture were published a few years later: 'The view which I am so bold to put forth considers radiation as a high species of vibration in the lines of force which are known to connect particles, and also masses of matter, together. It endeavours to dismiss the aether but not the vibrations.'[1]

Just how 'high' the number of vibrations occurring in light is was graphically illustrated a few years later in a delightful book by John Tyndall, another nineteenth-century scientist. Tyndall pointed out that the speed of light is so great that a 'column' of light 300,000 km long enters your eye every second, and the wavelength of (for example) red light is so short that there are, in round numbers, 200 thousand billion ripples in a column of red light that long, all interacting with the receptors in your eye every second to produce the sensation of sight[2].

[1] From Michael Faraday, *Experimental Researches in Electricity, Vol. II* (London: Taylor & Francis, 1855), p. 451.
[2] Tyndall, *On Light*, p. 63; I have modernized the numbers.

Faraday's own insight into the nature of those waves was vindicated over the next two decades by the work of Maxwell, who eventually described the vibrations of electric and magnetic fields in a set of four equations published in 1864, three years before Faraday died.

THE COLOURS OF MAGIC

Maxwell's background was very different from that of Faraday. He was the greatest theoretical physicist between the times of Newton and Einstein, and he came from a prominent eighteenth-century Scottish family, the Clerks of Penicuik. During the eighteenth century, there were two intermarriages between the Clerks and another wealthy family, the Maxwells of Middlebie; James's father, John Clerk, adopted the name Maxwell when he inherited the Middlebie estate, running to 1500 acres of farmland near Dalbeattie, in Galloway, in the south-west of Scotland. John Clerk Maxwell was a lawyer, but had a keen interest in science, and was a Fellow of the Royal Society of Edinburgh. So James had both a secure and comfortable family background, and an early introduction to the scientific world.

He was born in 1831 in Edinburgh, where his parents had stayed so that his mother could receive the best medical attention during her confinement. But he spent the first ten years of his life at Glenlair House, on the estate in Galloway. He was an only child, and his mother taught him to read and supervised his early education, but died of cancer, at the age of 48, when he was eight. Dalbeattie was still far out in the sticks in those days. Until the railway came to the town in 1846, it was a full day's journey to Glasgow, and two days to Edinburgh before the Glasgow–Edinburgh line opened in 1837. Maxwell had no close contemporaries, and on top of the trauma of his mother's death he was saddled for the next two years with a tutor who had decidedly old-fashioned ideas about education and the importance of learning Latin by rote. He was then sent to study at the Edinburgh Academy, living with one of his two Edinburgh aunts when in town, and returning home to Glenlair for holidays.

The initial impact made by the Galloway boy on his fellow pupils at the Academy was not that of a genius in the making. He had a country accent, different clothes from the city boys', and shoes that his father, demonstrating more practical skill than sound common sense, had designed and made himself. Maxwell returned to his aunt's home at the end of his first day at school bruised and battered, with torn clothes, and a new nickname, 'Dafty'. The name stuck throughout his time at the Academy, although it was clearly meant to indicate his strangeness, rather than any lack of intellect.

In spite of his inauspicious start, Maxwell got on well at the school. He showed his mathematical ability, a few years later, by devising a way to draw an oval (not an ellipse, as some biographies have it) using a looped piece of string. Because of his father's contacts with the scientific establishment in Edinburgh, the discovery was published, as Maxwell's first scientific paper, when he was 14. In all honesty, though, this was not a particularly world-shattering discovery – but it did give Maxwell an introduction to the scientific community in Edinburgh at an early age.

In 1847, when he was 16 (the usual age for university entry in Scotland at that time), Maxwell moved on to Edinburgh University. After completing three years of the four-year course, however, he left for Cambridge, graduating in mathematics in 1854. He became Professor of Natural Philosophy at Marischal College, in Aberdeen, in 1856, but when the College merged with King's College to become the University of Aberdeen, in 1860, one of the two Professors of Natural Philosophy had to step down. Maxwell lost his job, even though he was by now married to the daughter of the Principal of Marischal College, because he was the junior. He moved to King's College in London for five years, but returned to the family home in Scotland when his father died in 1865. He stayed there, as a gentleman farmer and amateur scientist, for six years (taking the opportunity to write up his great work on electricity and magnetism in book form), but was persuaded back to Cambridge in 1874 to become the first Professor of Experimental Physics at Cambridge University and first head of the Cavendish Laboratory, which he was instrumental in creating as one of the world's leading centres of scientific excellence. But he died in 1879 – at the same age as his mother had died, and of the same disease, cancer.

Maxwell's interests covered a broad spectrum of nineteenth-century physics, including the kinetic theory of gases, heat and thermodynamics, the nature and stability of the rings of Saturn, an accurate estimate of the sizes of molecules, and other contributions. But his seminal work was on the nature of light and colour. His first really dramatic discovery seemed more like magic than science, and showed how to make colour photographs from black-and-white images – a process still used today, among other things by space probes sending back colour pictures from Saturn and other distant parts of the Solar System. When such probes send back pictures of Saturn's rings, they are using a colour-photography technique invented by Maxwell, to obtain images of a ring system explained by Maxwell; and the signals are sent back to Earth as radio waves, part of the electromagnetic spectrum, whose properties were also explained (indeed, as far as radio is concerned, predicted) by Maxwell. Magic indeed!

Maxwell invented colour photography as a development from Young's idea that colour vision involves three kinds of receptors in the eye, each sensitive to just one of the primary colours, red, green and blue (among other things, Young's theory neatly explains colour-blindness, caused by the failure of one or more of the sets of receptors). Maxwell began studying the way different colours interact in 1849, when he was a student at Edinburgh University, working in the laboratory of James Forbes. Forbes already knew Maxwell before the young student came to the University – he was the Professor of Natural Philosophy there, and had been the person to whom Maxwell's father showed the paper about ovals, leading to its publication in the *Proceedings of the Royal Society of Edinburgh*. Forbes and Maxwell used spinning coloured disks in their investigations – the disk would be divided up into segments, each one a different colour, and then spun, to see what effect the mixture of colours had on the eye.

Forbes had to abandon his experiments because of a severe illness, and Maxwell soon left Edinburgh; but he took up the experiments again in 1854, after he graduated from Cambridge. He showed how different colours could be built up from mixtures of the three primary colours, and invented what he called a 'colour box', in which the three primary colours could first be separated out from sunlight and then recombined in different proportions to produce any desired colour.

The crowning achievement of this aspect of Maxwell's work came in 1861, when he projected the first colour photograph to a suitably impressed audience (which included Faraday) at the Royal Institution. This picture was the forerunner of all colour photography, and of the process used in colour TV. Maxwell took three separate photographic images of a piece of tartan ribbon, one each through a red, a blue and a green filter. Each filter transmits light of only one colour. So each photographic plate contained the information – the pattern of light and shade – corresponding to just one colour of light. But each photograph was simply a black-and-white image – three different black-and-white images of the same ribbon, each with a different emphasis of light and shade, but none of them containing a hint of colour.

The three photographic plates were then projected simultaneously onto a screen, carefully arranged for the projected images to overlap each other exactly, using the same filters in the respective projectors. The images could be projected one at a time, as red, green or blue images alone. This established that the combined image on the screen contained only a mixture of red, blue and green light; but it was exactly the same mixture as in the light reflected from the original ribbon. The combined image clearly showed all the colours of the tartan ribbon, proving that human perception of colour makes use of only the three primary colours.

A distant space probe does essentially the same trick by taking three pictures through three filters, and radioing the data (information about patterns of light and shade) from each image back to Earth, where it is reconstructed into a combined image by a computer. Your TV does the equivalent trick because the screen is covered with sets of three tiny dots, each dot capable of producing a spot of light with one of the three primary colours. The colour picture is produced by triggering the right combination of the three colours to make the right colourful pattern of light and shade across the screen.

But although Maxwell's demonstration to the Royal Institution was a triumph, it actually owed more to magic than he realized at the time. Although the eyewitness accounts of the event leave no doubt that a true-colour image was seen on the screen, years later photographers realized that the chemicals used in the photographic plates used in the demonstration could not have been sensitive to red light at all. The puzzle was resolved in the early 1960s by researchers at the Kodak Laboratories in the United States. They found that the red colour in Maxwell's tartan ribbon also reflected ultraviolet light (invisible to the human eye), and that by a pure coincidence the red filter he used also just happened to let ultraviolet light through. The pattern of light and shade on the 'red' plate in Maxwell's demonstration was actually the pattern produced by ultraviolet light, but thanks to this double coincidence it was exactly the pattern that would have been produced by red light if the plate had been sensitive to red light.

The original photographic plates used by Maxwell have been

preserved in Cambridge, and were used in a reconstruction of the original Royal Institution demonstration in 1961, a hundred years after Maxwell had carried it out. The three plates still gave a full-colour image of the tartan ribbon, even though by then all the observers knew that the 'red' component of the image had been obtained by a lucky fluke. Maxwell's magic, at least in this case, consisted not just of producing full-colour images using only black-and-white photographs and red, green and blue light, but of getting the right answer for (partly) the wrong reason. In his greatest contribution to science, however, he undoubtedly got the right answer for the right reason – but it was an answer which left a great deal for the next generation of physicists to puzzle over.

MAXWELL'S AMAZING EQUATIONS

Maxwell began his major investigation into electricity and magnetism just after he graduated from Cambridge University in 1854. In the previous decade, William Thomson, who lived from 1824 to 1907 and became Lord Kelvin in 1892, had found a mathematical analogy between the flow of heat through a solid and the pattern of electric forces in a region. Maxwell was intrigued, and looked for other analogies of this kind, tossing ideas backward and forward with Thomson in a series of letters. His first published work on electricity and magnetism, in the mid-1850s, developed the analogy between Faraday's lines of force and the 'streamlines' of an incompressible fluid.

The fact that electricity could be described by equations similar to those describing such diverse things as heat in a solid or the flow of a liquid should not, he said, be taken as an indication that electricity is really 'like' either of them – the analogies were purely mathematical, 'a similarity between relations, not a similarity between the things related'.[1] The same sort of equations describe the movement of heat, water and electricity – but that doesn't mean electricity 'is' water, any more than it means water 'is' heat.

Over the next ten years, Maxwell extended the analogy between

[1] Quoted in Everitt, *James Clerk Maxwell*, p. 88.

electricity and the flow of a fluid. He developed what now seems a bizarre physical image of the forces of electricity and magnetism being conveyed through the interactions of whirlpool-like vortices spinning in a fluid that filled all of the space between material objects – the ether. In some ways, this was a step backwards from Faraday's insight that there is no need for an ether, and that the forces themselves – the fields – are what matter. But the physical image Maxwell worked with in those years was not so important as the equations he was deriving. The same mathematics could describe different physical systems, as the analogies to water and heat show; and Maxwell's equations, whatever image they conjured up, did accurately describe the way forces between electrically charged or magnetized objects worked – *provided* the properties of the vortex medium (the ether) were chosen in the right way.

His next leap of imagination was to consider what would happen if the vortex medium could be squeezed and stretched – if it was elastic. Obviously, waves can travel through an elastic medium. The speed those waves move at depends on the properties of the medium, and in 1862 Maxwell found that, with the choice of properties needed to explain the forces of electricity and magnetism, the medium would transmit waves at the speed of light. His excitement shone through in the words of a paper published that year, in which the italics are Maxwell's own: 'We can scarcely avoid the inference that *light consists in the transverse undulations of the same medium which is the cause of electric and magnetic phenomena.*'[1]

There was still a lot of work to do to refine the mathematical description of electromagnetic phenomena and light. Maxwell found that he could abandon the concept of the vortex theory altogether, and explain every known electrical and magnetic phenomenon in terms of *A Dynamical Theory of the Electromagnetic Field*, the title of a paper published in 1864. The theory summed up everything there was to say about electricity and magnetism in a set of four equations, now known as Maxwell's equations. If you want to know the force with which two electric charges of a certain size, a certain distance apart, attract one another, the answer can be found by solving Maxwell's equations. If you want to know the strength of the electric current generated by a particular moving magnet, the answer can be found by solving Maxwell's equations. *Every* problem involving electricity and

[1] *On Physical Lines of Force*, quoted in Everitt, *James Clerk Maxwell*, p. 99.

magnetism (except for some quantum effects, discussed in the next chapter) can be solved by using Maxwell's equations; they represented the greatest scientific step forward since the time of Newton. And the equations contain a certain number, a constant always written as c, which corresponds to the speed with which electromagnetic waves move.

The value of c is found by experiments which measure the electric and magnetic properties of charges at rest or moving in wires. It is a number that comes out of studies of electricity and magnetism alone. As Maxwell himself said, 'the only use made of light in the experiment was to see the instruments'. But the number that comes out of the experiments as the value of c is exactly the speed of light:

> This velocity is so nearly that of light that it seems we have strong reason to conclude that light itself (including radiant heat and other radiations, if any) is an electromagnetic disturbance in the form of waves propagated through the electromagnetic field according to electromagnetic laws.[1]

Maxwell realized that there could be other forms of electromagnetic wave than those we see as visible light – heat, now known as infrared radiation, and 'other radiations', including what we know as radio waves. The prediction of other forms of electromagnetic radiation was confirmed in the 1880s, when Heinrich Hertz produced long-wavelength electromagnetic radiation, using electric currents moving up and down vertical wires, and measured its speed. These radio waves did indeed travel at the speed of light, as predicted by Maxwell, and, like light, they could be refracted, reflected and diffracted in appropriate experimental set-ups.

In the modern interpretation of Maxwell's equations, the ether and the vortices have vanished, and been replaced by the reality of Faraday's lines of force, the electromagnetic field. Of course, this is only the latest mental image to hold centre stage; we have no better idea of what is 'real' for an electron than Faraday, or Maxwell, or anyone else, had. The advantage of the field theory is that it is simple, and that it shows the way the mathematics works in a clear-cut way. But models should never be regarded as anything more than an aid to the imagination, a way of helping us to picture (or calculate) what is going on. The reality resides in the mathematical equations themselves,

---

[1] Both quotes from Maxwell's 1864 paper are taken from Baierlein, *Newton to Einstein*, p. 122.

whether the equations are used to describe electromagnetic waves, heat in a solid, or the flow of water. As long as the equations correctly tell us how the system will change when it is disturbed in a certain way, it doesn't really matter how you picture the interplay of forces at work.

But still, most people do need to have analogies and models in order to picture what is going on. The simplest mental image of the way light moves is to think in term of waves rippling along a rope. Remember that a moving magnetic field creates an electric field, and that a varying electric field creates a magnetic field. Think of two waves, moving together in step, like the ripples you can send along a stretched rope by jiggling one end. Let's say that the electric ripples move vertically, up and down along the rope; then, the magnetic ripples move sideways, from left to right, at right angles to the electric ripples. At any point along the rope, the strength of the electric field is constantly changing, as ripples pass through. The changing electric field produces a changing magnetic field. So at every point along the rope, the magnetic field is constantly changing as the ripples march through. The changing magnetic field produces a changing electric field. The two changing fields march hand in hand, each responsible for the other, as a beam of light, driven by the energy released from the source of that light.

That neat picture, though, was far from being established in 1864. As late as 1878, in an article for the *Encyclopaedia Britannica*, Maxwell himself still promoted the idea of an ether: 'Whatever difficulties we may have in forming a consistent idea of the constitution of the aether, there can be no doubt that the interplanetary and interstellar spaces are occupied by a material substance or body.'[1]

Maxwell's theory had gained wide support by the time of his death, but only really became established as 'the' theory of light with the investigation of radio waves a decade later. The work which spelled the death-knell of the ether (and which was partly inspired by that *Encyclopaedia Britannica* article of 1878) was also carried out in the 1880s, although the full significance of that work would not become clear until after the beginning of the twentieth century. As for the person who would explain to the world the real significance of the constant *c* in Maxwell's equations, and put those crucial experiments in perspective, he was not quite eight months old when Maxwell died in November 1879. His name was Albert Einstein, and his entry onto the stage signals the step forward into the modern times of physics.

[1] See, for example, Zajonc, *Catching the Light*, p. 146.

CHAPTER TWO

Modern Times

Isaac Newton knew about relativity of motion, and so did the physicists of the nineteenth century. The Moon in its orbit moves relative to the Earth, and the Earth moves relative to the Sun. If you drive a car at 50 kph down a straight road, and overtake me, riding my bicycle at 15 kph in the same direction, then you are moving at 35 kph relative to me. When Maxwell's equations provided a precise value for the speed of light, it was natural for physicists of the time to assume that this meant the speed of light relative to the ether, the substance which was thought to transmit light. Since the Earth moves in a roughly circular orbit around the Sun, however, the Earth cannot always be moving at the same speed relative to the ether. Sometimes it moves in one direction; six months later, on the opposite side of its orbit, it is moving in the opposite direction. A combination of Newtonian ideas about the relativity of motion and the idea of light as an electromagnetic wave transmitted through the ether led naturally to the conclusion that the speed of light *relative to the Earth* must be different at different times of year.

Some astronomers tried, unsuccessfully, to detect this difference in studies of the light arriving from the stars and planets at different times of year. But there is also a way to measure the effect using Earth-based experiments. If a beam of light moves in the same direction in which the Earth is moving, we ought to be trying to overtake it, so that it moves slightly less rapidly relative to our measuring instruments. But a beam of light moving across the direction of the Earth's motion, at right angles, should be measured to have the full speed c determined by Maxwell's equations.

Of course, the effect of the Earth's motion is small compared with the speed of light. Light moves at 300,000 km per second (in round terms) while the Earth's orbital speed is just under 30 km per second – in round terms, just 0.01 per cent of the speed of light. In his article on the ether for the *Encyclopaedia Britannica*, Maxwell pointed out how to measure the velocity of the Earth relative to the ether, using light itself to make the measurements. In principle, it would be possible to split a beam of light into two, and send each of the two beams on a 'journey' bouncing between a pair of mirrors. One beam would travel in the same direction as the Earth's motion in its orbit, while the other beam would travel between a pair of mirrors at right angles to the Earth's motion in its orbit. Then the two beams could be brought back together, and allowed to interfere, just like the light from the two holes in Young's experiment. The two beams should have been moving at slightly different speeds relative to the Earth. So, provided the experiment was carefully set up so that they both covered the same distance, they would have got out of step with one another, and would produce interference fringes. The spacing of the interference fringes would reveal exactly how fast the Earth was moving relative to the ether. But, Maxwell concluded, the effect would be so small that it would be impossible to detect. The challenge was, however, almost immediately taken up, by a young American researcher.

THE DEATH OF THE ETHER

Albert Michelson had actually been born in Germany, in 1852; but his family emigrated to the United States when he was still a child. He graduated from the US Naval Academy at Annapolis in 1873, and spent two years at sea before being appointed to a teaching post at the Academy. As an instructor in physics and chemistry, one of his duties was to demonstrate to the midshipmen at the Academy how the speed of light could be measured. Dissatisfied with the results of the standard experiment of the time, he set about improving it. To do so, he developed a more precise experiment. The skills he developed in this work meant that he was well placed to take up the challenge posed in Maxwell's *Britannica* article, using interferometry to measure the Earth's motion through the ether; this led to a lifetime's work developing ever better interferometers and using them to make increasingly precise measurements with the aid of beams of light.

The method Michelson used to measure the speed of light was based on bouncing light beams off rotating mirrors. The technique had been pioneered by the Frenchman Jean Foucault, who lived from 1819 to 1868 and also invented the gyroscope and demonstrated the rotation of the Earth, using his famous pendulum. His speed of light measurements involved bouncing beams of light off a flat mirror that was spinning very fast. The reflected beam then bounced off another mirror and back on to the first mirror, which had moved a little while the light beam was travelling to and from the second mirror. The angle by which the light beam was deflected, because of the rotation of the mirror, revealed how long the beam had taken to make the double journey to the second mirror and back.

Foucault used this technique to show for the first time, in 1850, that light travels more slowly in water than in air, confirming that it travels as a wave. In 1862, he had refined the technique sufficiently to measure the speed of light as 298,000 km per second, within one per cent of the best modern value.

Michelson refined the technique further by adding more mirrors and extending the length of the path travelled by the light beam. He used a rotating octagonal drum of mirrors (and later drums with more than eight faces) to deflect the light beam. With the drum rotating at a known speed, the eight mirrors would each briefly flick into position to make the reflection at precisely known intervals. By varying the speed of the drum to get the appropriate reflections, bouncing the light off one face of the drum on its outward journey and another face of the drum on its return, Michelson could tell how long the light had spent on its journey.

In the ultimate version of this experiment, carried out in 1926 when Michelson was 73, the journey the light beam went on was a 70-km-long round trip between two mountain peaks in California; the figure Michelson came up with in 1926 for the speed of light was 299,796 ± 4 km per second. Within the range of experimental uncertainties, this matches the accepted modern value, 299,792.5 km per second. When asked why he had bothered, at his age, to make the effort to measure c so precisely, he replied 'because it is such good fun.'[1] Michelson died in 1931, in his 79th year, while having fun planning an even more accurate measurement of the speed of light.

In the early 1890s, Michelson had also, with his colleague Edward Morley, measured the length of the standard metre in Paris in terms of the wavelength of pure light from the red part of the spectrum;

---

[1] Quoted in Weber, *Pioneers in Science*, p. 33.

they were way ahead of their time, but in 1960 essentially the same technique was officially adopted to define the length of the metre in terms of the properties of light. For his pioneering efforts along these lines, his measurements of the speed of light and his skill at constructing precise optical instruments, Michelson became the first American to be awarded the Nobel Prize, receiving the physics award in 1907. But his name is best remembered today for the experiment he carried out with Morley in the second half of the 1880s.

In 1880, Michelson had left Annapolis on what was supposed to be a temporary study leave, going off to Europe to work in Berlin, Heidelberg and Paris. He had, of course, read Maxwell's *Britannica* article on the ether, and it was in 1881, while working at Hermann Helmholtz's laboratory in Berlin, that he first tried to measure the motion of the Earth relative to the ether. He used the technique suggested by Maxwell and an interferometer of his own design, built with the aid of funds provided by Alexander Graham Bell. But he found no evidence of the predicted effect. Nobody was too concerned at the time, however, because the experiment was difficult (and might therefore be wrong), and because it had been suggested that the Earth might drag the ether along with it, so that measurements made on the surface of the Earth would not detect any 'ether drift'.

Michelson never did return to work at Annapolis, but resigned from the Navy, and went to be Professor of Physics at the Case School of Applied Science, in Cleveland, Ohio, in 1882. One of the first things he did there was to measure the speed of light, coming up with a figure of 186,320 miles per second (299,845 km per second), which was the most accurate measurement then made, and remained so for ten years, until Michelson himself improved on it.

In 1885, however, the Dutch physicist Hendrik Lorentz showed that the ether drag effect would not work, and the astronomical measurements were incompatible with the idea that light travels at a fixed speed relative to the ether, while the Earth moves through the ether. This encouraged Michelson to join forces with Edward Morley, who was Professor of Chemistry at the college which became Western Reserve University (merging with Case), also in Cleveland.

Like Michelson, Morley (who lived from 1838 to 1923) devoted his career to precise measurements, including the oxygen content of the air, and the atomic weight of oxygen. Combining his skills with Michelson's, the team built an improved version of the interferometer experiment, and tried once again to measure the motion of the Earth through the ether. In 1887, they confirmed Michelson's original result, but with such precision that there was no room left to hope that there really might be something going on, but that the instruments were not sensitive enough to detect it. There was no evidence at all that the Earth moves relative to the ether. Or, to put it another way, the speed of light is exactly the same, whichever way it is moving relative to the Earth.

How could this be?

TOWARDS A SPECIAL THEORY OF RELATIVITY

On reflection, it's just as well that there is no evidence that the ether exists. For, when you think about it, it turns out that the kind of ether the Victorians believed in would have to have a very peculiar combination of properties. In the first place, it has to be extremely stiff, in order for light waves to move through it so rapidly. Vibrations passing through a substance move more rapidly if the substance is stiffer – the speed of sound is greater in a steel rod than it is in air, for example. But the speed of sound in air is just 344 metres per second, and even in steel it is only 5000 metres per second. Try to imagine a substance so stiff that vibrations travel through it at 300,000 *kilometres* per second, and you have some idea of one of the key properties of the ether.

On the other hand, the ether must be very tenuous. After all, the Earth moves through the ether, seemingly unimpeded – it isn't slowed down in its orbit by the drag of the ether. And the ether was supposed to be everywhere, in order to propagate light – even between the atoms and molecules of air itself. You would be wading through the ether every time you took a step, and breathing it in by the lungful, without it having any effect on you at all except to transmit light from one place to another.

Even without the work of Michelson and Morley, perhaps before too long nineteenth-century scientists would have decided that the idea of the ether should be discarded after all. The alternative proposal put forward by Faraday, that the electric and magnetic fields of force extend through empty space, was still not fully accepted a generation later, even after Maxwell's equations had shown how varying electric and magnetic fields could propagate hand in hand as an electromagnetic wave. But its time was about to come.

The first signs of just how dramatic a change in the physicists' conception of the world would be required to explain the behaviour of light came soon after Michelson and Morley reported their definitive experimental results in 1887. The Irish physicist George Fitzgerald, who had been born in Dublin in 1851, had already made a mark in science by correctly predicting the way in which an oscillating electric current should produce what we now know as radio waves, pointing the way to the work of Heinrich Hertz. Now, in 1889, he offered an explanation of the results of the Michelson–Morley experiment. The failure of the experiment to measure any change in the speed of light, regardless of which way the light is moving relative to the Earth, could be because the entire experimental apparatus (and, indeed, the Earth itself) shrinks in the direction of motion. This would solve the problem – the speed of light relative to the Earth was 'really' dependent on the Earth's motion through the ether, on this picture, but the measuring equipment had shrunk by just the required amount to give the illusion that the speed was still c.

This was not a completely off-the-wall idea. Physicists already knew – indeed, it had been shown by Maxwell – that the force between two electric charges depends on how they are moving. A stronger force would pull things together more tightly, and what Fitzgerald was suggesting was that the forces holding atoms and molecules together would be stronger if they were moving (again, at this stage the implicit assumption is still *moving relative to the ether*), pulling them closer together and shrinking anything they were made of.

The same idea was put forward independently by Hendrik Lorentz in the 1890s; somewhat unfairly, I always feel, it is known as the Lorentz–Fitzgerald contraction, rather than the Fitzgerald–Lorentz contraction. But Lorentz, who lived from 1853 to 1928, and received the 1902 Nobel Prize in Physics for his work on electromagnetism, did take the idea much further than Fitzgerald, and in 1904 (three years after Fitzgerald died) he developed a set of equations known as the Lorentz transformations, which describe how not only the length of a moving object but also its other properties 'transform' when viewed by observers moving at different velocities.

In fact, Lorentz developed his transformation equations to describe mathematically the way electromagnetic fields would look to different observers; the transformations put the relative velocities of the observers into Maxwell's equations. It was Albert Einstein who, a year later, showed that the same transformation equations apply to mechanical systems, and described the way not only length but time, velocities and even the masses of moving objects look different for observers moving at different velocities. Curiously, however, although Einstein used Lorentz's work on electromagnetism as a jumping-off point, when he developed his special theory of relativity he was not influenced by the evidence from the Michelson–Morley experiment that the speed of light is always the same. Near the end of his life, responding to an enquiry made in 1954, a year before he died, Einstein said that that experiment was 'not a considerable influence. I even do not remember if I knew of it at all when I wrote my first [1905] paper on the subject.'[1] So what *did* start him thinking along the lines that were to revolutionize physics in the first decade of the twentieth century?

EINSTEIN'S INSIGHT

In 1905, Einstein was 26 years old. He had graduated from the Technology Institute (ETH) in Zurich in 1900, and since 1902 he had been working as a technical expert in the Swiss patent office in Bern, assessing the technical merits (or otherwise) of new inventions. His ambitions for an academic career seemed, at the time, to have been dashed by his failure to take the conventional education offered by the ETH entirely seriously. Although he did reasonably well in his final examinations, he had a reputation for laziness and had alienated some of the professors who might have found a position for him. But the work in the patent office was easy, and gave him time to develop his ideas on physics – sufficient time to publish several scientific papers and complete his PhD thesis in the years leading up to his breakthrough with the special theory of relativity.

Einstein's life and subsequent achievements would fill several books (and have);[2] but here I want to concentrate on the special theory, and what it tells us about the nature of light. Einstein's great gift was his

[1] Quoted in Weber, *Pioneers of Science*, p. 33.
[2] My own contribution to the Einstein industry, written with Michael White, is listed in the Bibliography.

genius for physical insight into what a problem was really all about. Mathematics was never his strongest point, although he was certainly stronger at maths than most people, but he had a great feel for physics. The insight which led him to the special theory was based on his sound physical intuition about what Maxwell's equations were really saying. He puzzled over what would happen if you could ride alongside a beam of light, at the same speed the light was moving.

Remember that the nub of Maxwell's equations is that a *changing* electric field produces the (changing) magnetic part of the wave, and the *changing* magnetic field produces the (changing) electric part of the wave. But if you were moving at the same speed as the wave, it would not be 'waving', from your point of view, at all. It would be stationary, like a wave on the sea frozen into ice before it could break. And Maxwell's equations quite clearly said (and, of course, experiments had shown) that a stationary magnetic field would not make an electric field, and nor would a stationary electric field make a magnetic field. There would simply be no wave at all – not even a frozen one.

Once again, the problem comes back to relativity of motion. Newton himself, although he realized that motion is relative as far as people moving about on the Earth, or birds flying through the air, or boats sailing on the sea are concerned, also thought that there must be some ultimate frame of reference against which all motion could be measured – a universal standard of rest. The idea of the ether fitted this picture, with all motion measured relative to the ether. Newton also believed that there was an absolute standard of time, a kind of clock of God, which ticked away inexorably at the same rate for everybody. But these seemingly sensible ideas could not be made to square with Maxwell's equations.

Einstein saw that there is no need to invoke a preferred frame of reference at all. There does not have to be a standard of rest in the Universe against which all velocities are measured. Instead, he said that *all* motion is relative – which means that anybody is entitled to say that they are at rest, and to measure all motion relative to themselves. Strictly speaking, this relativity of motion applies only to observers moving at constant velocities relative to one another – that is, at constant speeds and in straight lines. Anyone in an accelerated frame of reference can tell they are moving from the forces they feel, such as the way your weight seems to change as a fast elevator starts or stops, and the way you are flung to the side of a vehicle going round a bend at high speed. It is this restriction that gives the theory the name 'special'. Einstein's general theory of relativity extended the idea to cover acceleration, motion along curved trajectories, and gravity; happily, though, we don't need the general theory at all for the discussion in this book.

As for the electromagnetic waves that make up a beam of light, they do not know, or care, about the speed with which the source of the waves is moving; once they are up and running, they progress through space at the speed c determined by Maxwell's equations.

If all observers moving at constant velocities (all inertial observers, in the jargon of physics) are entitled to say that they are at rest and measure all motion relative to themselves, it follows that they must all find the laws of physics to be the same. If I carry out an experiment in my spaceship, travelling at three-quarters of the speed of light relative to the Earth, I must get the same 'answers' as you get in your spaceship, travelling at half the speed of light relative to the Earth. If we got different answers, we would know which of us was 'really' moving and which one wasn't.

So how do you have to modify Newton's description of reality to ensure that all inertial observers get the same results when they carry out physics experiments? Einstein found the answer by thinking about how a pulse of electromagnetic radiation spreading out from a light source would look to observers moving at different velocities. In the frame of reference of the light source, the light forms a spherical shell spreading out through space. So it must look like a spherical shell to *all* inertial observers, or they would know that they were moving. The only way in which the shell of light can look spherical to all inertial observers is if their measuring rods are shrunk by their motion relative to the source of the light. The shrinkage is exactly the Lorentz–Fitzgerald contraction, calculated using the Lorentz transformations. But there is more – in particular, velocities themselves do not add up in the way they ought to if common-sense Newtonian ideas applied.

Newtonian common sense would say, for example, that if you see a spaceship fly past you at a speed three-quarters that of light ($0.75c$), and another spaceship moving in the opposite direction, also at $0.75c$, then the speed of one spacecraft relative to the other one must be $1.5c$. But according to the Lorentz transformations, an observer in either spaceship will measure the velocity of the other one as $0.96c$. What's more, if the occupant of either spaceship flashes a light, then the occupants of both spaceships will measure the speed of the electromagnetic waves in that light pulse as c, not $1.75c$. In fact, there is no way of adding up two velocities smaller than *c*, using the Lorentz transformations, to get an answer as big as *c*, let alone bigger. Among other things, this means that if you start out moving slower than *c* and go faster and faster (adding more and more to your speed) you will never be able to get up to *c*. You can always go faster relative to some chosen frame of reference – from 0.9*c* to 0.99*c*, from 0.99*c* to 0.999*c*, and so on – but you can never match the speed of light (and always, when you measure the speed of light itself, you get the answer *c* relative to yourself!).

It's worth taking this on board again, slowly, because it is one of the essential features of the best resolution of the quantum mysteries:

> The special theory of relativity tells us that it is *impossible* to run alongside a beam of light at the same speed as the light is moving; relative to some chosen inertial frame, you can in principle get your own velocity up as close to the speed of light as you like without actually reaching it – but no matter how close you get, when you measure the speed of the light beam itself you will always get the answer *c*.

There are many fascinating implications and repercussions of the special theory of relativity, which I do not have room to go into in detail here. This is the theory that tells us, for example, that mass and energy are connected, by the famous equation $E = mc^2$; it is the theory that unites space and time into one whole, spacetime. But the one thing that is relevant for the present discussion is that the special theory tells us that time runs more slowly for a moving clock. There is no God-given absolute time that applies to all observers.

This time-dilation effect is governed by the same Lorentz transformation equations as the Lorentz–Fitzgerald contraction. One way of getting a picture of what is going on is, indeed, to think in terms of spacetime, rather than space or time alone. Hermann Minkowski, who had actually been one of Einstein's teachers at the ETH, came up with the idea in 1908, and said that time should literally be regarded as a fourth dimension, setting 'forward/backward' in time on the same footing as 'forward/backward' in space, 'up/down' in space and 'left/right' in space. The one key difference is that time enters the relevant equations with the opposite sign to the space dimensions – conventionally, they have a ' + ' sign and time has a ' − ' sign, although the equations would work just as well the other way around. As a result, while motion makes lengths *shrink*, it makes time intervals *expand*. The two effects are matched with one another, so that the amount by which a moving object shrinks is exactly balanced by the amount by which time expands for it.

Relativists describe objects as having a kind of four-dimensional length, which they call extension, and which stays the same, no matter how the object moves. Depending on how it does move, though (or how the observer moves relative to the object), the extension seems to be divided up in different ways between length and time.

You can see something similar happening in three dimensions if you hold a pencil under a light and look at the shadow that it casts on the floor. Depending on how you orient the pencil, its shadow could be any length, from nothing at all to the actual length of the pencil, even though its real length stays the same. Moving at constant velocity through three dimensions is mathematically equivalent to changing the orientation of an object in four-dimensional spacetime, and the changing length of the shadow is equivalent to the changing amount of length contraction the object undergoes, while the amount of time dilation goes the opposite way, increasing as the shadow shrinks. The three-dimensional world around us is essentially a shadow from four-dimensional spacetime.

None of these effects shows up until the velocities involved are a sizeable fraction of the speed of light. But one of the most important points to note is that they *do* show up, and exactly in the way Einstein's theory predicts. The special theory has been tested in very many experiments, and has passed every test with flying colours; I will give you just one classic example of time dilation at work.

The atmosphere of the Earth is constantly being bombarded by particles from space, called cosmic rays. When these particles interact with atoms high in the atmosphere, they often produce showers of another kind of particle, called muons. These muons have a very short lifetime. They exist as muons for only a couple of microseconds, before they 'decay' into other kinds of particle. Even though they travel at a large fraction of the speed of light, they do not live long enough, according to everyday common-sense ideas about time, to get through the atmosphere to the surface of the Earth. And yet, particle physicists find that most of these muons do get down to the ground. The explanation is that because the muons are moving so fast relative to the Earth, time is running more slowly for them. To be precise, the special theory of relativity says that the lifetime of the muons is extended by a factor of 9 – they live 9 times longer, according to our clocks, than they would if they were sitting still.

But remember that the special theory also says that the muons are entitled to regard themselves as sitting still. In their own frame of reference, surely they should still decay before reaching the ground? Not at all! If the muons are regarded as being at rest, which is indeed allowed, then we have to regard the Earth as rushing past the muons at a sizeable fraction of the speed of light! This, of course, will cause the Earth to shrink, from the point of view of the muons, by the amount calculated from the Lorentz transformations. Because the speed involved is the same, and because of the symmetry between space and time in those equations, the amount of the contraction is the same as the amount of the time dilation – a factor of 9. But because of the opposite sign in front of the time part of those equations, the thickness of the Earth's atmosphere *shrinks* by a factor of 9. From the muons' point of view, the distance they have to cover is only one-ninth of the distance we measure for the thickness of the Earth's atmosphere, and they have ample time to complete such a short journey before they decay.

The special theory of relativity is not just some crazy hypothesis, but passes Newton's experimental test – it 'explains the properties of things' *and* 'furnishes experiments' which can be used to test (successfully) those explanations.

So what happens when we push this time-dilation business to the limit? Getting back to the original question that Einstein asked about light, how does the Universe 'look' to a beam of light (or a photon, if you prefer), or to a person riding on a light beam? And how does time flow for a photon?

To answer the second question first – it doesn't. The Lorentz transformations tell us that time stands still for an object moving at the speed of light. From the point of view of the photon, of course, it is everything else that is rushing past at the speed of light. And under such extreme conditions, the Lorentz–Fitzgerald contraction reduces the distances between all objects to zero. You can either say that time does not exist for an electromagnetic wave, so that it is everywhere along its path (everywhere in the Universe) at once; or you can say that distance does not exist for an electromagnetic wave, so that it 'touches' everything in the Universe at once.

This is an enormously important idea, which I have never seen given due attention. From the point of view of a photon, it takes no time at all to cross the 150 million km from the Sun to the Earth (or to cross the entire Universe), for the simple reason that this space interval does not exist for the photon. Physicists seem to ignore this remarkable state of affairs, because they know that no material object can ever be accelerated to the speed of light, so no human (or mechanical) observer is ever going to experience this strange phenomenon. Perhaps they are simply so stunned by what the equations say that they have not fully thought out the implications. As I hope to persuade you, though, this curious behaviour of space and time from the point of view of photons may help to resolve *all* the outstanding mysteries of quantum physics. But before I begin to show you how the special theory of relativity and the quantum theory have already been combined to provide an up-to-date description of electromagnetic phenomena, it's worth looking briefly at another implication of the special theory. Einstein's equations tell us that you can never get a relative velocity greater than that of light by adding up two velocities (or more!) that are less than *c*. But they do *not* say that it is impossible to travel faster than light.

FASTER THAN LIGHT/BACKWARDS IN TIME

As I hinted in the Prologue, the special theory of relativity does not say that it is impossible in principle for something to travel faster than light. What it does say is that it is impossible to cross the speed-of-light 'barrier'. If a particle is moving slower than light, then it would take an infinite amount of energy even to accelerate it up to the speed of light. But Einstein's equations have a beautiful symmetry, in the way they describe motion, with light speed in the middle. They also say that if a particle did exist that travelled faster than light, it would *always* travel faster than light. On the other side of the light barrier, it would require an infinite amount of energy to *slow down* the particle to light speed.

Because the equations allow for the possible existence of such faster-than-light particles, they have even been given a name, tachyons, from the Greek word meaning *swift* (slightly tongue in cheek, a few physicists say that ordinary slower-than-light particles also deserve a name, and call them tardons, because they are 'tardy' compared with tachyons). If tachyons do exist, they inhabit a very strange world which in a sense 'mirrors' the laws of physics that we know. The symmetry of the equations relative to the speed of light means that this critical velocity seems, in a sense, to repel particles on either side of it. It is like an infinitely long, infinitely high mountain ridge; particles on our side of the ridge roll down the slope to slower speeds, if left to their own devices, but particles on the other side of the slope roll down to higher speeds, unless they are given an energy boost. And since time runs more and more slowly as you approach the speed of light ('climbing the ridge') from our side, coming to a halt at light speed itself, it should be no real surprise to discover that time runs slowly *backwards* just the other side of the barrier, and more and more rapidly backwards for tachyons moving more and more rapidly as they 'descend the ridge' on the other side – as they move further away from light speed.

As a tachyon loses energy, it goes faster both through space and (backwards) through time. So the fate of any tachyon created in a particle interaction (perhaps when a cosmic ray interacts with the Earth's atmosphere) is to radiate away all its energy in a brief burst, while accelerating up to a fantastic speed and scooting off to the other side of the Universe.

It is extremely unlikely that such entities really do exist. But even the faintest possibility of discovering something so exciting is worth devoting a little attention to, in the same way that you might think it worth buying a lottery ticket in order to have a small chance of winning a large prize. So some physicists have actually looked for traces of tachyons in cosmic-ray showers (this represents a very modest 'bet' indeed, since the detectors are already built and being used for more conventional work). Logically enough, the 'signature' of a tachyon would be an event recorded in a cosmic-ray detector on the surface of the Earth just *before* a shower of particles such as muons was created by the impact of a particle from space at the top of the Earth's atmosphere. Any tachyon created in that event would travel backwards in time on its way down to the detector!

Unfortunately for science-fiction fans (and for the physicists, who would surely win the Nobel Prize if they caught a tachyon), there is no good evidence from these experiments that tachyons do exist. The importance of the idea of tachyons is still simply that it demonstrates how the equations of relativity theory do not rule out the possibility of entities that travel backwards in time. Nobody suggests that material particles – tachyons – are created when the intelligent alien opens the spacecraft and notices whether one kitten is dead or alive, and these particles then travel backwards in time to collapse the 'original' electron wave function (apart from anything else, making particles, even tachyons, requires energy, in the form of mc^2). But if the laws of physics permit any kind of backwards-in-time communication, we have to be willing to extend our ideas about what is happening to the spacefaring kittens in this direction, as well as considering the possibility of action at a distance.

As I spelled out in my book *In Search of the Edge of Time*, there is actually nothing in the laws of physics (including those of the general theory of relativity, not just the special theory) that forbids time travel. It would be extremely difficult, and it runs counter to common sense. But it is not forbidden by the laws of physics, and our common-sense ideas have already taken a battering from both relativity theory and quantum theory, each backed up by experiment in the way Newton would have approved of.

I won't labour the point. But tuck it away in the back of your mind for later. Then, some of the things I have to say at the end of this book won't come as quite such a shock to you. For now, let's get back to light itself, and in particular to the link between electromagnetism and quantum physics.

ENTER THE PHOTON

By the end of the nineteenth century, the idea that light is a form of wave was so firmly established that it would have been almost heretical to suggest that it could behave as a particle. Yet it turned out that that was just what had to be suggested to explain the behaviour of light. It took until the 1920s, though, for physicists to come to terms (insofar as they ever have come to terms) with the idea of photons, and the wave–particle duality.

The first step was taken by Max Planck, a German physicist of the old school, who had been born in 1858 and by 1892 had become Professor of Physics at the Institute for Theoretical Physics in Berlin. In the second half of the 1890s, Planck made a heroic effort to explain the way electromagnetic radiation, including light, is radiated by hot objects. Like other physicists of the time, he was confronting a huge puzzle. According to the classical laws of wave behaviour – laws which work superbly well when applied to the vibrations of a guitar string, or to waves moving on the surface of a pond – it should be easier for charged particles to radiate energy at higher frequencies (which correspond to shorter wavelengths). Inside a hot object (like the filament of a light bulb) there are charged particles (electrons) jiggling about at a speed which depends on their temperature. Any hot object, according to the classical picture, should therefore radiate intensely in the short-wavelength part of the spectrum (ultraviolet, X-rays and so on), and very little at longer wavelengths (in the visible, infrared and radio bands). But your light bulb certainly does not radiate copious quantities of X-rays, or you wouldn't be alive to be reading these words. In fact, any hot object radiates most strongly at a band of wavelengths centred on a characteristic wavelength that depends on its temperature. The Sun is yellow because it has a temperature of roughly 6000 degrees, and yellow is the colour most strongly radiated at that temperature; a red hot poker is slightly cooler than the Sun, and so radiates most strongly at slightly longer wavelengths, in the red part of the spectrum. The link between temperature and the characteristic wavelength of radiation is known as the black-body law, and the characteristic radiation is known as black-body radiation ('black'-body because the same rules apply to the way radiation is absorbed by a black surface; once again, there is a symmetry in the equations).

After a great deal of work, involving trips up several blind alleys, Planck found a way out of the dilemma in 1900. He realized that the nature of black-body radiation could be explained if it was not, in fact, possible for hot objects to radiate any amount of electromagnetic energy that they liked. Instead, the electromagnetic energy is emitted (or absorbed, depending on which way round you run the equations) in packets of a definite size, which he called quanta. Each packet of waves has an energy which depends on its frequency (the energy is actually equal to the frequency multiplied by that certain number now known as Planck's constant). This explains the nature of black-body radiation, as follows.

Although the speed with which the electrons in a hot object vibrate depends on the temperature, they do not all move at exactly the same speed. Most oscillate at about some average rate, but some will have a bit more energy, and vibrate more quickly, while some will have a bit less energy, and oscillate more slowly. There is always a spread of energies around the average value, just as there is always a spread of heights around the average height of a class of schoolchildren. For very high frequencies, the energy needed to make one quantum is correspondingly large, and very few of the charged particles in a hot object (the oscillating electrons) will have that much energy available to make a quantum. So only a few short-wavelength quanta are emitted. At the other extreme, for quanta with low energy there are many electrons able to make the corresponding radiation, but the energy involved is so feeble that even all the long-wavelength quanta added together don't count for much. But in the middle, at a range of frequencies corresponding to the temperature of the body, there will be plenty of oscillating electrons able to make the quanta, and plenty of energy in each quantum to add up to give an impressive glow.

The announcement of this discovery by Planck in December 1900 is seen as the beginning of the quantum revolution. But Planck himself did *not* say that light could only exist in the form of quanta, as little particles of light. He thought that the important point was that some property of the charged particles that do the radiating (or absorbing) of electromagnetic energy was at work, and that although light itself (and other forms of electromagnetic radiation) existed as a classical wave, the properties of charged particles prevented them emitting or absorbing radiation except in definite amounts.

Although Planck's calculations gave all the right answers when they were applied to describing electromagnetic radiation from hot objects, many people (including Planck himself) were unhappy about how to use them to interpret what was 'really' going on. It wasn't until 1918 that Planck received the Nobel Prize for his work (ironically, he never did come to terms with the new quantum theory, although he lived until 1947). The timing owed a great deal to the theoretical work of Albert Einstein (he received his own Nobel Prize for this work, in 1921), and the experiments of Robert Millikan, who received his Prize in 1923.

Einstein alone had the courage, at the beginning of the twentieth century, to accept the physical reality of Planck's quanta. He explained, in a paper published in 1905, the way in which electrons are knocked out of a metal surface by light (the photoelectric effect) as due to the impact of particles of light (quanta) with the electrons in the metal. Each quantum carries a definite amount of energy, which depends only on its frequency (colour). So for pure light of a particular colour all the electrons knocked out of the metal carry the same energy.

Experimenters had been puzzling over this discovery since 1899, and now there was an explanation. Einstein was well aware of the revolutionary nature of his discovery. Hardly anyone took the idea seriously at first, and even in 1911, at a scientific meeting known as the First Solvay Congress, he told his colleagues: 'I insist on the provisional character of this concept, which does not seem reconcilable with the experimentally verified consequences of the wave theory.'[1] The problem was that even Einstein was still thinking in either/or terms. *Either* light was a wave, *or* it was a particle. Evidence for waves must rule out the possibility of particles; evidence for particles must rule out the possibility of waves. They couldn't *both* be right, could they?

Millikan, who lived from 1868 to 1953, and was working in the University of Chicago at the time of the First Solvay Congress, agreed with that sentiment; he thought that the suggestion that light could be made of particles was nonsense, and set out to prove Einstein was wrong in a superbly designed and executed series of experiments on the photoelectric effect. By 1915, against what he had thought was his better judgement, he was forced to admit that all the evidence showed that Einstein was right, and that light quanta did have a real, physical existence. Along the way, he obtained the first accurate measurement of the size of Planck's constant; he also measured the charge on the electron with great precision. Still, nobody understood the significance of the physical reality of light quanta; but the experimental evidence could not be denied, and the flurry of Nobel Prizes related to the work, starting with Planck, soon followed. By 1923, when Millikan received his prize, the idea of light quanta was firmly established; but they were only given the name 'photons' (from the Greek word for light, *photos*) in 1926, by Gilbert Lewis, a physicist based in Berkeley, California. This name followed hot on the heels of the discovery of a new way of describing the behaviour of particles of light, which itself led to the creation of quantum mechanics itself.

---

[1] See Gribbin, *In Search of Schrödinger's Cat*, p. 81 *et seq.*

THE MAN WHO TAUGHT EINSTEIN TO COUNT PHOTONS

By showing physicists that one and one do not necessarily add up to make two, the Indian physicist Satyendra Nath Bose, working at the University of Dacca in what was then East Bengal, paved the way for quantum mechanics and a theory of light and matter. The year 1994 marked a triple anniversary in the life and times of Satyendra Bose. He was born in Calcutta exactly 100 years before, on 1 January 1894, and he died there just 80 years later, on 4 February 1974. His greatest achievement, in the early 1920s, was to take the rag-bag of ideas that comprised the quantum theory of radiation at that time and to provide a mathematical description of light quanta which tied everything into a coherent whole.

When Planck introduced the idea of quantization into the discussion of how radiation and matter interact, at the end of the nineteenth century, he had used the idea in an *ad hoc* way to explain the behaviour of black-body radiation. Even though Albert Einstein had suggested in 1905 that light itself must be quantized (and Millikan's experiments had shown that he was right), even in the early 1920s many physicists – perhaps most – did not 'really believe' that light existed in the form of particles. It is no coincidence that the particle of light was only given a name, the 'photon', in 1926, after Bose put the quantum theory of light on a secure mathematical footing.

Planck had solved the black-body problem by cutting the electromagnetic energy up (mathematically) into small chunks. But I stress that he did not suggest that these pieces of radiation had any physical significance, but thought that what was happening inside a hot object to make it radiate energy only allowed energy to be radiated in pieces of a certain size. This is rather like the way in which water drips from a tap into a slowly filling sink. Inside the pipes behind the tap, there is a continuum of water as an amorphous liquid, and in the sink there is an amorphous pool of water; but the physical properties of the leaky tap mean that water only escapes from the tap in droplets of a certain size.

Like the emission of water from the dripping tap, in Planck's description of black-body radiation, it was only the mechanism of emission (or absorption) of radiation that involved droplets of a certain size. There was no suggestion, not even from Planck himself, that light, or other forms of electromagnetic radiation, really did only exist in the form of little lumps, or quanta. In a letter written to R. V. Wood in 1931, Planck recalled that '[quantization] was purely a formal assumption and I really did not give it much thought except that no matter what the cost, I must bring about a positive result'.[1] In the early 1920s, almost everyone knew that 'the light quantum' could explain otherwise puzzling features of the interaction between light and matter, but hardly anyone believed that this was more than a mathematical trick; they still thought of light as 'really' being a wave, described by Maxwell's equations.

There was, however, one exception. In India, physicists took the light quantum seriously. The pioneering astrophysicist Meghnad Saha used the light quantum to describe radiation pressure in a paper published in the *Astrophysical Journal* in 1919, and then collaborated with Bose to produce one of the earliest English translations of Einstein's papers on the general theory of relativity. This led to discussions which made Bose aware of the need for a proper derivation of Planck's 'law' of black-body radiation, free from the inconsistencies which were the inevitable consequence of the way Planck had grafted the essential ingredient of quantum discreteness onto a classical framework of continuous waves. He found that this could be achieved – *provided* that the particles of light obeyed a different kind of statistics from those we are used to.

The curious thing about Bose's work was that it did not include any vestige of a description of electromagnetic radiation in terms of waves, or indeed of electromagnetism. He arrived at Planck's equation by treating the photons that fill a cavity as a gas of particles, obeying a different kind of statistical law from the kind of statistics used in the everyday world.

The simplest way to get a picture of what is going on is to think of a pair of newly-minted coins, of the same value. If you toss both coins, there are three different outcomes that you might see. There might be two heads showing, or two tails, or one of each. At first sight, you might guess that each outcome is equally probable – that there is a one-in-three chance, for instance, of getting a head–tail combination. But a little thought shows that this is not the case.

Suppose you were to mark one of the coins in some way, so that the two coins are distinguishable (or use two coins of different denominations). Now it is easy to see that although there is only one

[1] Quoted by Dipankar Home, *New Scientist*, 8 January 1994.

way to get the combination head–head, and one way to get the combination tail–tail, there are *two* ways to get the combination head–tail (think of these as being 'head–tail' and 'tail-head'). *Either* coin can be 'head' provided the other one comes up tails. So the right way to count the possible results of tossing two coins is as *four* possible outcomes, head–head, tail–tail, head–tail and tail–head. The chance of any one outcome is one in four (a quarter), not one in three. And since there are two ways of getting one head and one tail, the chances of this pattern turning up are one in two, or 50 per cent (a quarter plus a quarter). The important point is that if the coins are indistinguishable then the combination head–tail cannot be distinguished from the combination tail–head.

But if the particles really are distinguishable from one another (not because you have marked the coin, but because of their intrinsic properties), the statistics are different. Then there are four different outcomes to the coin-tossing experiment, each equally probable. Don't worry too much about the details; all that matters is that you can see from this simple example that the statistics are indeed different if particles are indistinguishable and if they can be told apart. In other words, the statistics you have to use to describe the behaviour of large numbers of particles acting together depends on what kind of particle you are dealing with.

Bose found that he could derive Planck's formula by treating photons as particles which have to be counted in a certain way. Photons are indistinguishable from one another (although this is not the whole story, I do not intend to go into all the complications here), and in the photon world the statistical way in which the photons behave affects the way in which energy is shared out among them – the distribution of photons among different energy states.

There are other intriguing features of the behaviour of photons. They are not *conserved*. You make more photons, for example, every time you flick a switch to turn on a light, and they also stream out from the Sun and stars in vast numbers. Photons are constantly being absorbed by the walls of your room, by your eyes, by the surface of the Earth, and so on. But these two processes are not in balance, and the number of photons in the Universe is constantly changing.

This is quite different from the behaviour of the kind of particles we are used to thinking of as particles, such as electrons. Electrons cannot be created nor destroyed, except in special circumstances where an electron and its 'antiparticle' counterpart, a positron, are created (or destroyed) together. The total number of electrons in the Universe (for this purpose, a positron counts as 'minus one' electrons) stays the same.

It turns out that a different kind of statistics apply to particles such as electrons; these statistics are known to quantum physicists as 'Fermi–Dirac' statistics, in recognition of the work of the Italian Enrico Fermi and the Englishman Paul Dirac. Particles that obey what are now known as Bose–Einstein statistics, such as photons, are collectively known as 'bosons'; particles that obey Fermi–Dirac statistics, such as electrons, are collectively known as 'fermions'.

Why 'Bose–Einstein', and not just 'Bose' statistics? In 1924, Bose sent a paper describing his discoveries to the *Philosophical Magazine*, but received no response. So in June that year he sent a copy of the paper to Einstein. He asked Einstein to read the manuscript (it was written in English) and, if he thought it made sense, to pass it on for publication in the *Zeitschrift für Physik*. Einstein was so impressed by the work that he translated it into German himself, and submitted it to the journal with his endorsement. Anything endorsed by Einstein was certain of a welcome at the *Zeitschrift*, and the paper duly appeared in print in the summer of 1924.

The implications were awesome. Bose had derived the black-body equation for electromagnetic radiation simply by treating photons as real particles obeying a certain kind of statistics and behaving as a quantum gas. There is *no vestige* of an electromagnetic wave in Bose's derivation of the black-body law! Einstein himself took up the idea of the new statistics, and applied it to other problems in three papers that were his last major contributions to quantum theory. Using the new statistics to describe the behaviour of gases under different conditions (the statistics can apply, in some cases, to entities that are conserved), among other things he showed that just as the behaviour of light (traditionally regarded as a wave) could be explained in terms of particles, so, under the right circumstances, molecules ('particles') ought to behave as waves.

Just at the time when he was puzzling over the significance of this discovery, late in 1924, he was sent a copy of the PhD dissertation of Louis de Broglie by de Broglie's supervisor, Paul Langevin. De Broglie had made the seemingly outrageous claim that particles such as electrons could behave as waves. Langevin could not decide if it was a stroke of genius, or completely crazy. 'I believe,' Einstein wrote, 'that it involves more than a mere analogy.' De Broglie's work was taken seriously on the strength of this seal of approval, and was taken up by Erwin Schrödinger, who developed it into a complete description of the quantum world, wave mechanics. He later remarked that 'wave mechanics was born in statistics', and said in a letter to Einstein in April 1926, 'The whole thing would not have started at present or at any other time (I mean as far as I am concerned) had not your second paper on the Bose gas directed my attention to the importance of de Broglie's ideas.'[1]

Bose himself, however, did not take part in the exciting development of quantum theory over the next few years. Instead, he followed up his other early interest, in the general theory of relativity, following Einstein up what turned out to be the blind alley of a premature search for a unified field theory. After Einstein died in 1955, this line of research lost its thrust, and Bose's contributions are largely forgotten. For the last 20 years of his life, he devoted himself to the popularization of science, teaching, and improving the public understanding of science. 'I was not really in science any more,' he commented late in his life. 'I was like a comet, a comet which came once and never returned again.' But the blazing light shed by that comet changed the way physicists thought in the 1920s, and changed the way physics has progressed ever since.

It would be more than 20 years after photons were named before physicists finally came up with a satisfactory theory of the quantized electromagnetic field – but the wait would be worth it, because the theory they did eventually come up with, known as quantum electrodynamics (QED for short), is the most successful and accurate scientific theory there has ever been. It tells how electrons and electromagnetic radiation interact with one another, it explains everything in the physical world except gravity and the behaviour of atomic nuclei, and it has been tested to an extraordinary degree of precision in experiments.

[1] Quotes in this and the next paragraph from Dipankar Home, *New Scientist*, 8 January 1994.

THE STRANGE THEORY OF LIGHT AND MATTER

I have borrowed the title of this section from the subtitle of Richard Feynman's superb book *QED*. Feynman, who was born in 1918 and died in 1988, was the greatest theoretical physicist of his generation, who made many important contributions to science, wrote both a best-selling textbook and best-selling volumes of autobiographical memoirs, was a highly regarded and popular teacher, and by the end of his life one of the most famous scientists on Earth (and certainly *the* most famous scientific 'character'). But of all of his many achievements the greatest[1] was undoubtedly QED, what he called 'the strange theory of light and matter'.

QED is so important because the interactions of electrons with one another and with electromagnetic radiation determine almost everything about the world around us. That world, and ourselves, are made of atoms, and atoms consist of a compact central nucleus surrounded by a cloud of electrons. The electrons are the visible 'face' of the atom, and interactions between atoms and molecules are really interactions between the electron clouds. The way electrons interact is by exchanging photons. An electron emits a photon, and 'recoils' in some way, or an electron absorbs a photon, and gets a 'kick'. Everything that happens when atoms interact can be explained in these terms.

All of chemistry is explained by quantum physics, and specifically by QED; biological life depends upon the behaviour of complex molecules such as proteins and DNA, which is also chemistry and also depends, ultimately, on the quantum properties of electrons. The way electrons are held in a cloud around the nucleus of an atom depends on the interaction between the negative electric charge of the electrons and the positive electric charge of the protons in the nucleus, so it is also governed by QED. Things like radioactive decay, which involve changes in the atomic nuclei themselves, cannot be explained by QED, but require a different theory. But even our best current understanding of what goes on inside atomic nuclei is based on theories

---

[1] At least, the greatest in scientific terms. When a colleague of mine, Marcus Chown, was a student at CalTech he asked Feynman to explain to his (Chown's) mother why physics was important. Feynman wrote to her to put things in perspective. He told her not to worry about what her son's work was all about. 'Physics is not important,' said Feynman in that letter, 'love is.'

which are deliberately modelled on the success of QED itself, and which are highly successful in their own right, although not as successful as QED.

There are different ways to explain what QED is all about, but I like the way Feynman explains it. He deals in terms of particles – photons and electrons – and probability waves. The probabilities tell you where you are most likely to find the particles, but when you do find them (as in the electron version of the experiment with two holes) you do indeed find them as particles. There are only three things that matter in working out how light and matter interact, says Feynman. First, there is the probability of a photon going from one place to another. Secondly, there is the probability of an electron going from one place to another. Thirdly, there is the probability that an electron either absorbs or emits a photon. If you can calculate all of the event probabilities involving all the electrons and photons, you can work out what will happen when electrons and photons interact.

For complicated systems, this would involve an enormous number of calculations, even though each individual calculation might be quite simple. So precise calculations have only been carried out for relatively simple systems involving a few electrons exchanging a few photons. Nevertheless, these precise examples help to establish more general approximations (which can still be pretty accurate) that apply in more complicated situations.

Part of the complication in the calculations is that when I glibly mention 'the probability of a photon (or an electron) going from one place to another place' the image conjured up in your mind is, almost certainly, of a particle moving along a smooth trajectory from A to B. But this is wrong! One of Feynman's key contributions to the development of QED was his realization that we have to take account of every possible path from A to B. We have already seen, from the experiment with two holes, how a single photon passing through the experiment seems to be aware of both holes, as if it had followed both routes through the experiment. But Feynman goes further. In going from one place to another, he says, a particle takes account of every possible route. Not just straight-line routes, or smoothly curving routes, but the most complicated and wiggly 'trajectories' that you can imagine.

The notion sounds ludicrous at first; but the way Feynman came up with the idea shows that it is (almost!) common sense. In the experiment with two holes, the probability of getting a spot of light

at a particular point on the screen beyond the slits has to be worked
out by adding up the probabilities corresponding to the light passing
through each of the two holes. This really is common sense, as long
as we don't worry about the particle aspect of the nature of light. But
suppose we made four slits in the obstructing screen, instead of two.

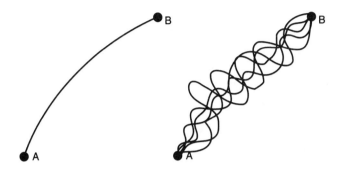

Figure 7

*Classical physics – the physics of Isaac Newton – says that a particle follows a unique
trajectory from A to B. Richard Feynman's version of quantum mechanics says that we
must calculate the effect of all the possible paths from A to B and add them together – not
just the few paths shown here, but literally every possible path. This 'sum-over-histories'
(or 'path-integral') approach is one way of understanding how a single electron can go
through both holes at once in the experiment with two holes and interfere with itself.*

Then, we would have to add up four sets of probabilities. With eight
holes, we have to add up eight lots of probability, and so on. If the
obstructing screen were sliced with a million razor slits, we could still,
in principle, work out the brightness of the light on any part of the
far screen by calculating all the probabilities for the million different
paths through the experiment and adding them up ('integrating' them).
By then, there would be more holes than obstruction in the screen.
But why stop there? We can imagine dividing the obstructing screen
up into more and more slits, until eventually there is no obstruction
left – the 'slits' all overlap with one another. With no screen left,
Feynman realized, we have to add up the probabilities for every
possible route from the source of the light to the spot on the far
screen, and if there is no obstructing screen at all that means integrating
probabilities for literally every conceivable path through the experiment. The probabilities for the more complicated routes are very
small, and they usually cancel each other out of the calculation. But

their influence is really there, as he explains by describing how light reflects from a mirror.

One of the things we all learn in school is that light bounces off a

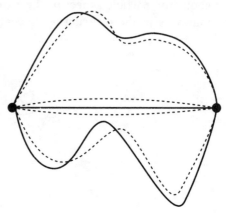

Figure 8
Feynman's approach to quantum mechanics also applies to light. Light doesn't really travel only in straight lines, but by every possible path from a source to an observer. It just happens that when you add up the 'histories' they all cancel out except for the ones close to a straight line.

mirror in such a way that the angle it makes to the surface coming in is the same as the angle it makes going out. You can easily check that this is so by looking into a mirror at an angle, and noticing which objects in the reflection are in your line of sight. This is an example of another thing we learned in school or college, that light travels by the route that takes the least time. Allowing for the fact that the light is being reflected, and not heading straight from the source to your eyes, this equal-angle reflection is indeed the path which makes the total distance from the source to your eyes least, and which therefore minimizes the time it takes the light to get to you. You would be amazed if you were told that the light from an object actually travels to every part of the mirror on its way to your eyes, and that reflections coming from everywhere on the mirror, at all kinds of crazy angles, combine to make the image you see. Well, prepare to be amazed – that is exactly what does happen, according to the laws of quantum physics.

At one extreme, imagine light from the object going towards the mirror at right angles, then bouncing off at a shallower angle to reach your eyes; alternatively, think of light travelling at a shallow angle to

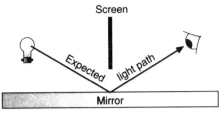

Figure 9
Classical physics says that a mirror reflects light in straight lines so that the angle of incidence is always equal to the angle of reflection.

hit the mirror right in front of your eyes, then bouncing out at right angles so that you see it. Indeed, imagine the light going off in the opposite direction, *away* from you, to the far edge of the mirror, then bouncing back at a sharp angle to your eyes. All these, and all other possibilities, really are happening. The reason we don't notice this is that the paths near to the shortest-distance path are not only more likely, but reinforce one another to make the shortest-distance path

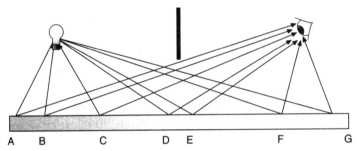

Figure 10
Feynman discovered that even though straight-line paths are favoured by the sum-over-histories, all straight-line paths are important. Light reflects from all parts of the mirror, at all kinds of crazy angles. But this time the paths from adjacent strips of the mirror cancel each other out, except near the classical path.

overwhelmingly more probable. The reason is that it is only near to this 'classical' trajectory that the probabilities add up, and reinforce one another. As Feynman puts it, 'where the time is least is also where the time for nearby paths is nearly the same',[1] and that is where the probabilities add up. Out on the edges of the mirror, where the photons are bouncing off at crazy angles in order to get to your eyes, there is a much bigger difference in the time it takes for 'neighbouring'

[1] Feynman, *QED*, p. 45.

photons to get from the source to the mirror and up to your eyes. The way that the probabilities work out, this means that the probabilities for neighbouring paths almost cancel each other out. So only the part of the mirror that you instinctively know, from common sense, is import-ant really matters in making the reflection.

But wait. This isn't the end of the story. There is a simple experiment to show that reflected photons really are arriving from way out on the edge of the mirror, even though they are cancelling out. Imagine covering up all of the mirror, except for a piece out by the edge, with a black cloth, so that it cannot reflect. Now, it is no use looking for an image where the light rays would be if they travelled down to the black cloth and bounced off at an equal angle to reach your eyes, because the light is absorbed by the cloth. You won't see any image at all. There is, though, a trick, to make an image using only the edge of the mirror, at the 'wrong' angle to make a reflec-tion.

Although the probabilities for *neighbouring* parts of the edge of the mirror cancel out, you can still find strips of mirror where the probabilities add together. The problem is just that they are separated by strips of mirror where the probabilities go the other way. Overall, the probabilities cancel; but there are strips of reinforcing probab-ility separated by strips of opposing probability. All we have to do is lay some more strips of black cloth carefully over the regions where the probability is 'wrong', and we will be left with half as much visible mirror, but with all the probabilities reinforcing each other.

The spacing of the strips needed to make the reinforcement work depends on the wavelength of the light involved (a beautiful example of wave–particle duality, since we are describing the light beam in terms of photons!), so it is best to do the experiment with light of one colour (monochromatic light) if you want to see a clean image. But when you do it, the experiment works. You take a piece of mirror in the wrong place to get a reflection, and sure enough you don't get a reflection. Then you cover up half the mirror in the right way (so that common sense would say there is even less chance of seeing a reflection), and you do get a reflection.

Such a system of reflecting strips is called a diffraction grating (because the reflection effect can also be explained in terms of diffraction of light waves), and you've probably seen exactly this effect at work many times. A 'grating' with the reflecting strips at a set distance

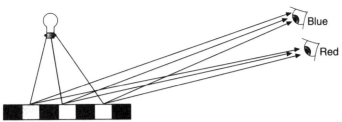

Figure 11
You can actually see the light reflecting at all kinds of crazy angles if you black out
parallel strips of the mirror. This means that the 'cancelling' light paths are removed. By
covering up half the mirror, you can actually get more *reflection! Different colours of light*
reflect at slightly different angles from such a diffraction grating, to produce a rainbow
effect. The trick only works for very narrow strips of mirror and blackout, but if you hold
a compact disc under a bright light at an angle you can see it for yourself.

apart will reflect different colours of light at slightly different angles, spreading out white light to make the same rainbow-pattern spectrum Newton saw in sunlight that had been spread out into its colours by a prism. And this is what makes rainbow patterns when you hold a compact disc under the light. Hold the disc at the correct angle, the one corresponding to common sense, and you will see an ordinary reflection of the light bulb in its shiny mirror surface; but when you tilt the disc, so that you can no longer see the common-sense reflection, you still see the rainbow pattern produced by photons bouncing off the parallel grooves in the mirror surface of the CD at crazy angles. Indeed, you can usually see bands of colour coming from light reflected from the 'wrong' part of the CD even when you can also see the 'normal' image. You can *see* QED at work in the privacy of your own home.

In this example, I have still only talked about light travelling in straight lines, bouncing off the mirror at different angles. In fact, the full version of the theory takes account of every possible way light can travel from one place to another, including crazy wiggly trajectories. Because the calculations involve adding up (integrating) all possible paths, this kind of approach to quantum physics is often called the 'path-integral' (or 'sum-over-histories') approach. Happily, though, the probabilities always add up so that it appears as if the light is travelling in a straight line. But the complete cancellation only happens away from the 'classical' straight line. 'Light doesn't *really* travel only in a straight line,' says Feynman, 'it "smells" the neighboring paths around it, and uses a small core of nearby space.'[1]

Similar descriptions of the way the probabilities add up explain everything in optics, including the way lenses work, the way light bends and slows down as it moves from air into water, the experiment with two holes, and Poisson's spot. But the triumph of QED is best seen by looking at how accurately it describes the way photons and electrons interact.

THE TRIUMPH OF QED

The simplest interaction is when an electron moving from one place to another either emits or absorbs a photon, and as a result ends up in a third place. The photon itself may have been emitted or absorbed by another electron, on another trajectory. Or it might, for example, be a photon associated with the magnetic field of a bar magnet. As long ago as 1929, Paul Dirac, one of the pioneers of quantum physics, developed a description of the way photons and electrons interact that took full account of the special theory of relativity, but didn't quite take full account of the requirements of quantum theory. In this description, Dirac, in effect, worked out the probabilities for an interaction between an electron and a photon, and used these to calculate a number which is a measure of how an electron interacts with a magnetic field (the property involved is called the magnetic moment of an electron). Dirac found that the number should have the value 1, in a certain system of units. But experiments showed that the number is really about 1.00116.

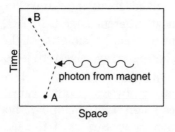

Figure 12
Paul Dirac's original calculation of the magnetic moment of an electron is based on a simple interaction involving one photon.

---

[1] Feynman, *QED*, p. 54.

The difference was small, but enough to show that the theory was incomplete. In 1948, Julian Schwinger, working at Harvard University, found a way to improve Dirac's calculation. Schwinger (who, incidentally, was born in the same year as Feynman, 1918, and the same city, New York) realized that while an electron was on its way from one place to another there was nothing to stop it emitting a photon of its own, and then reabsorbing the photon. This introduces a complication into the probability calculations, and the result is to make

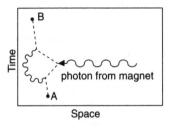

Figure 13
A more accurate calculation of the magnetic moment of an electron takes account of the possibility that the electron may also emit a photon and reabsorb it. Successive improvements in the calculations add more and more photons into the loop.

the calculated magnetic moment of the electron a little bit bigger. Not exactly the right size to match the experimental measurement, but a step in the right direction.

Once physicists had realized what was going on, the further steps needed to calculate the magnetic moment even more accurately were obvious, but they involved a lot of painstaking calculation. First, consider the possibility that the solitary electron, interacting with just one photon from the magnetic field, has actually emitted two photons of its own, one after another, and reabsorbed them. You have to consider every possible way in which this might happen, and add up the probabilities. The situation is already so complicated that it actually took two years for all the probabilities to be worked out and added up, but the result was even better agreement between theory and experiment.

By the middle of the 1980s, the calculations had been extended to include the effects of three 'extra' photons. Each complication is less probable than the one before, and makes a smaller correction to the calculation (and each complication is *much* harder to calculate than the one before). At this level the theoretical calculation of the magnetic moment is 1.00115965246, with an uncertainty of about 20 in the last two digits. Over the same decades, experimenters had improved their measurements of the magnetic moment, and were quoting a figure of

1.00115965221, with an uncertainty of about 4 in the last digit. The accuracy of these numbers is as good as measuring the distance from Los Angeles to New York, more than 3000 miles (about 5000 km), to the width of a human hair. And the agreement between the two numbers is the measure of how accurate QED is – the most precise and accurate theory, in terms of making predictions that are tested by experiment, that there has ever been. You may think it is crazy; you may not like it. But you cannot deny that it works – that the world really does operate in this way. In Feynman's words, 'nearly all the vast apparent variety in Nature results from the monotony of repeatedly changing just these three basic actions': the movement of a photon from one place to another, the movement of an electron from one place to another, and the interaction of an electron with a photon.[1]

But although the theory is so well borne out by experiments, it contains some very bizarre features – even more bizarre than I have yet let on. When two electrons interact by exchanging a photon, for example, that is precisely the correct way to describe what is going on. From the everyday point of view, it may seem that one electron emits a photon and that a little while later (or a long while later) the other electron absorbs it. But we are equally entitled to say that the second electron emits the photon 'in the future' and that it travels backwards in time, being absorbed by the other electron 'in the past'. This is not too wild an idea to grasp, especially since we have already learned that time has no meaning for a photon. But the same thing applies to the electrons themselves!

If a photon has enough energy, it can actually turn itself into a *pair* of electron-like particles (to do the trick, the E in the photon must be more than the mc^2 in two electrons). One of these particles is an everyday electron; the other is just like an electron but has positive charge instead of negative charge, and is called a positron. As ever, the equations that describe the process are symmetrical. When an electron and a positron meet they reverse the process and annihilate each other, to form an energetic photon. In a standard scenario, observed many times in experiments, an energetic photon moving from one place to another may turn into a positron–electron pair in this way. The two particles go off in different directions, and very soon the positron meets another electron and annihilates, producing another energetic photon.

[1] Feynman, *QED*, p. 110.

But, Feynman realized, this whole interaction can be described in terms of just one electron. This electron, on its way from one place to another place, interacts with an energetic photon. The interaction sends the electron travelling backwards in time until it interacts with another energetic photon, when it is 'turned around' again and travels forward in time. At each of the two interactions, there seem to be three entities involved – a positron, an electron and a photon. In the same way, where a ray of light bounces off a mirror there seem to be three entities involved – two rays of light making the appropriate angle where they meet at a point with the mirror, and the mirror itself. But just as there is really only one light ray, bounced back in space, so there is only one electron, bounced back in time. Photons can act as 'time mirrors' for electrons.

During the interval in which the electron is travelling backwards in time it looks to us, as we travel forwards in time, like a positron ('taking negative electricity away', by sending it back through time, acts like a classic double negative, and is exactly equivalent to adding positive electricity moving forwards in time). And as the calculations of the magnetic moment of the electron proceed to ever more complicated levels, you even have to take account of interactions like these occurring among the 'extra' photons associated with the electron.

This is very nearly everything I have to tell you about QED, and you probably think that it is more than enough. But I want to stress that these bizarre implications are not optional extras dreamed up to frighten people with. They are a fundamental feature of the best theory we have in physics, one that earned *three* people the Nobel Prize in 1965 (Feynman, Schwinger and the Japanese researcher Sinitiro Tomonaga) and is the jewel in the crown of science. You can't get rid of curiosities such as photons and even electrons travelling backwards in time without getting rid of QED itself.

That said, I also want to make one confession. There is a problem with QED. It is not quite the perfect theory. The problem basically involves what happens to an electron all by itself, going from one place to another. Even a lone electron can emit and reabsorb photons, and even these temporary photons can split into electron–positron pairs, which then annihilate one another to make the photon that is reabsorbed. Even these temporary electrons and positrons can emit further photons, provided they reabsorb them, and so on, and on. There is layer upon layer of complex interactions of this kind going on around every single electron, according to QED. The problem is that all of these possible self-interactions result in an endless addition of probabilities, so that the calculations of such simple properties as the charge or the mass of the electron blow up in our faces. The answers come out as infinity, which is clearly nonsense.

Feynman, Schwinger and Tomonaga found ways to get rid of the infinities. The trick is called renormalization, and in effect it involves dividing both sides of an equation by infinity to get the answer you want – something you were surely told at school is not allowed. The trick only works as well as it does because we *do* know, from experiments, the answers we want to get for the properties of the electron. Physicists accept renormalization because they have no choice – it can be made to give the right answers, whereas no other theory can be made to do so even by something as tricky as renormalization. So the three researchers were given the Nobel Prize for showing everyone how to do it. But a few years before he died Feynman said that 'having to resort to such hocus-pocus has prevented us from proving that the theory of quantum electrodynamics is mathematically self-consistent ... [renormalization] is what I would call a dippy process!'[1]

So QED in its present form is almost certainly not the last word, and there is still some work for the next generation of theoretical physicists to do. Nevertheless, any improvement on QED will still have to explain all the things that QED does, even more accurately, or it won't really be an improvement. Which means that we are stuck with path integrals, particles that 'smell out' neighbouring paths as they move through space, and particles that can be described, entirely consistently within the laws of physics, as moving backwards through time. Which brings me to another, much less heralded, discovery made by Feynman, one which may be the key to unlocking the mysteries of the quantum world.

LIGHT OF FUTURE DAYS

This was, in fact, the first of Feynman's many strikingly original contributions to physics; he made it in 1940, while he was a graduate student at Princeton, working under the supervision of John Wheeler.

---

[1] Feynman, *QED*, p. 128.

The infinities that plagued quantum theory were already a well-known problem at the time – the trick of renormalization would not be discovered for another eight years – and Feynman wondered if there might be a way to get rid of them by forbidding the electron to interact with itself. Unfortunately, that trick would not work.

When electrons are accelerated – given a push – they resist. They resist more than an uncharged particle resists being pushed. The electrons in an electric current moving in a wire will radiate away energy (in the form of radio waves) if they are accelerated, but they do not radiate as much energy as the amount that goes into pushing them through the wire. This is an extra form of resistance (called radiation resistance, because it is associated with the accelerations that produce radiation), over and above the ordinary electrical resistance of a wire to a steady current flowing through it. Radiation resistance only occurs because the electron is interacting with something; since it seemed that it could not be interacting with empty space, in the 1930s it was explained in terms of the electron interacting with itself, more or less in the way I have just outlined.

But Feynman had a bright idea. Nobody has ever seen a genuinely isolated electron, since there are vast numbers of all kinds of particles in the Universe (indeed, if there is anyone there to 'see' it, the electron is not isolated!). He imagined a Universe completely empty except for a single electron, and wondered whether it really could radiate electromagnetic energy at all. Perhaps, he suggested to Wheeler, you had to have a minimum of two electrons, one to emit the radiation and one to absorb it, before the radiation itself could exist. In a Universe containing just two electrons, the first could vibrate to and fro and radiate photons, while the second absorbed the photons and vibrated in its turn as a result, generating more photons that travelled over to the first electron and pushed back on it, providing the resistance to its original oscillation.

In this simple form, the idea would not work. The basic problem was that there would be a time delay – photons would have to travel from the first electron to the second one and back again before it 'noticed' any resistance to its vibrations. But, as we have seen, when photons are exchanged the direction of time does not come into the discussion. Jumping ahead of the present aspect of the story, QED (which hadn't been invented in 1940) does not distinguish between forwards and backwards in time, at least with regard to photons. This is logical, because QED is a relativistic theory that takes full account of the special theory of relativity, which says that time does not exist for a photon. If it takes zero time for a photon to be exchanged, it really doesn't matter on the photon's 'clocks' whether that is $+0$ or -0. The success of both theories confirms that Nature itself cannot distinguish between a photon moving forwards in time (from our point of view) and a photon moving backwards in time. All that Nature 'knows' is that a photon is exchanged.

Although QED hadn't been invented in 1940, Wheeler and Feynman knew that Maxwell's equations are themselves completely symmetric as far as time is concerned. When you solve the equations to describe the way a wave propagates, you always get two answers, one corresponding to a wave moving forwards in time and the other to a wave moving backwards in time. Again, with hindsight, this makes sense if light itself travels in zero time; but until Feynman came up with his new idea about how electrons radiate energy everyone had just ignored the second set of solutions to Maxwell's equations, since 'obviously' you could not really have waves moving backwards in time.

This, though, was just what Feynman and Wheeler needed to save his idea. Let's stick with the wave description of light for the rest of this chapter. Waves moving outwards from an electron, or a radio antenna, are called 'retarded' waves, because they arrive somewhere else after they have been emitted. Waves travelling backwards in time are called 'advanced' waves, because they arrive somewhere *before* they have been emitted somewhere else. You can think of retarded waves as like ripples spreading out evenly in all directions from a radio antenna, like the ripples spreading out from the point in a pond where a stone has been dropped; then, from our human perspective, advanced waves are like ripples moving smoothly in towards the antenna from all directions, like ripples starting from the edges of a pond and moving smoothly together in its centre. The analogy breaks down because all the energy in the advanced waves has nowhere to go when it gets to the centre of the pond; but advanced waves arriving at an electron from the Universe at large turn out to be just what you need to provide the drag we call radiation resistance. The incoming wave's energy is absorbed and opposes the motion of the electron. But how do the advanced waves know where to find the electron? Because the electron itself has told them where to look.

In the revised version of what came to be known as the 'Wheeler–Feynman absorber theory' (supervisors have a way of getting their names first on joint work with students), when an electron jiggles about it sends out both a retarded wave into the future and an advanced wave into the past. Wherever in the Universe (in space and time) this wave meets another electron (strictly speaking, whenever it meets any charged particle), it makes the other electron jiggle about. This jiggling means that the other electron also radiates, both into the future and into the past. The result is an overlapping sea of interacting electromagnetic waves, filling the entire Universe, as a result of a single electron jiggling about. Most of the waves cancel out, just as the probabilities largely cancel out in the quantum description of reflection. But some of those waves, from both past and future, return to the original electron, and provide the resistance needed to explain observations of the way accelerated electrons behave.

Early in 1941, Wheeler told Feynman to prepare a talk on the theory for the physics department at Princeton. This would be the young scientist's first formal presentation to such an audience, and, Princeton being Princeton and the year being 1941, the audience, although all 'in-house' physicists, would include Albert Einstein, Wolfgang Pauli (a pioneering quantum physicist so clever that in 1919, at the age of 19 and when still a student, he had produced a monograph on *both* the special and the general theories of relativity that was regarded as a model of clarity), and others who were lesser scientists only in comparison with these geniuses. After the talk, Pauli objected, mildly, that the description was really a kind of mathematical tautology, and asked Einstein whether he agreed. 'No,' said Einstein, 'the theory seems possible...'[1]

To say that Feynman never looked back would be an exaggeration, but no student can ever have had a more impressive seal of approval for his first real piece of research. So why was Einstein so impressed?

After what you have learned about path integrals, it should have come as no surprise to discover that when the calculations are carried through most of the complexity of the web of interacting waves cancels out, to leave a fairly straightforward 'reaction' on the original electron. None of the advanced waves survives in a form that would be detectable in any other way than through this reaction, and all that we can 'see' are the familiar retarded waves.

The great beauty about this, though, is that as far as the original electron is concerned the reaction is instantaneous. Some of it comes as a result of waves from the electron travelling into the future and

---

[1] Quoted in Gleick, *Genius*, p. 115.

generating waves which travel back into the past to arrive at the right time; some of it comes from waves that travel into the past and generate waves which then travel back to the future. But in every case, since according to a clock sitting next to the electron (or, indeed, any other clock) the time spent going forwards in time is the same as the time spent going backwards in time, the distance the waves have travelled does not matter. The reaction occurs as soon as the electron is accelerated. Wheeler–Feynman theory *can* explain radiation resistance, although it does not do the thing Feynman set out to do in the first place, remove the infinities from quantum theory. That is often the way in science; one problem may be the inspiration for a piece of research, but the research may end up solving a completely different problem (or posing previously unsuspected problems).

There is one further twist in the tale, which seemed, half a century ago, to be a fatal flaw in the theory. The whole business only works if every bit of electromagnetic energy radiated from an electron is 'reflected' in time in this way. If some of the radiation escapes into empty space and never meets another charged particle, the equations will not balance. It used to be thought that our Universe is infinite in extent, and 'open'. Trying to reflect all the radiation back in time to its origin would be like trying to trap the radiation in a box without a lid. The Wheeler–Feynman theory only gives the right answers if the Universe is like a closed box (or like the inside of a black hole) from which no energy can escape. And, would you believe, during the 1980s and 1990s, for reasons which have nothing to do with the Wheeler–Feynman theory, astronomers have come up with compelling evidence that the Universe is indeed 'closed'.[1]

Today, there is no conflict between the absorber theory and cosmology. Some theorists even suggest an intimate link between the fact that the Universe is expanding today and the way in which we only perceive the retarded wave, moving into the future, not the advanced wave, converging on all the charged particles. The Wheeler–Feynman idea stands as the best explanation of why radiation resistance occurs and how photons are exchanged between charged particles, although you would never know it from the way physics is taught in most colleges and universities. Curiously, this means that in a sense the ancients were right – your eyes *do* emit photons, as part of an exchange with the photons radiated by a source of light; but like the paths

---

[1] See my book *In the Beginning*.

involving photons bouncing at crazy angles off a mirror, they do not show up in the everyday world because of the way the probabilities cancel out. The important point we come back to, yet again, is that the old picture of a photon moving from a source of light to our eyes (or to anywhere else) is incomplete; time has no meaning for a photon, and all we can say is that photons have been exchanged between the source of the light and our eyes (or whatever).

You think this is strange? Everything I have described in this chapter is not only true, but very well established as a long-standing feature of physics. In a few years, the special theory of relativity will be a hundred years old; even QED is coming up to its fiftieth birthday. This is bedrock science, copper-bottomed and thoroughly understood (in terms, at least, of how to do the calculations), and confirmed time and time again by experiments. But if we really want to find an interpretation of quantum physics that gives us a feel for how the world really operates – what reality itself really is – we have to explain many more strange things. Some are old ideas that have only just been put to the experimental test; some are new ideas that have yet to be put to the test in experiments. All are very strange, and all are true.

CHAPTER THREE

Strange But True

One of the strangest features of the quantum world in general, and the behaviour of light in particular, is shown by the phenomenon known as polarization. At first sight, polarization seems to be a simple attribute of moving waves, and the explanation of polarization in those terms was one of the early triumphs of Maxwell's theory. Imagine, once again, holding one end of a stretched rope, with the other end tied to a tree. As before, by flicking your hand up and down, you can send ripples moving along the rope, caused by the rope moving up and down; this can now be thought of as a 'vertically polarized' wave. By flicking your hand from side to side, you can send similar ripples moving along the rope, but now they are caused by the rope moving from side to side; they are 'horizontally polarized' waves.

It is rather harder to get a feel for how a system consisting of two wave components at right angles to one another (the electric and the magnetic parts of the wave) can be polarized in this way, and the rope analogy breaks down completely when we think in terms of photons. But what matters is that even individual photons can carry a preferred orientation along with them. For lack of any better picture to carry in your mind, you can imagine that there is a pointer, or arrow, on each photon. The pointer may be vertical (for vertically polarized light), or horizontal (for horizontally polarized light), or anywhere in between (for light polarized at an intermediate angle).

Ordinary light, from the Sun or from an ordinary electric bulb, is not polarized. You can think of this as meaning that the arrows on all the myriads of photons streaming out from the light source are oriented at random, some one way, some another, with no preferred direction. But it is easy to make polarized light, by passing it through a substance which only allows photons with certain orientations to pass. Mixing the metaphor a little (and why not, since Nature herself seems to mix things up?), if the rope stretching from your hand to the tree passes through one of the slots in a tall picket fence, you will still be able to send vertically polarized ripples all the way along the rope to the tree, moving up and down as they pass through the slot; but any horizontally polarized ripples you initiate will be stopped when they get to the fence, because the rope cannot move to and fro horizontally without hitting the fence posts.

There are naturally occurring polarizing materials, notably the crystals known as calcite, and you can picture something similar to the rope-and-fence situation occurring when light waves meet the ordered ranks of atoms in such a crystal. Artificial polarizers, such as Polaroid sunglasses, are commonplace today. The sunglasses are effective for two reasons. First, because they only allow photons with one particular orientation to pass through, they cut out all the other photons and reduce the brightness of the light arriving at your eyes. Secondly, because light that is reflected from horizontal surfaces tends to be horizontally polarized, if the glasses are made to transmit only vertically polarized light (which they are), then they will cut out most of the reflected glare. This is why Polaroid glasses can be useful when driving at night (cutting down the dazzle from the lights of oncoming vehicles reflected off the road), as well as when the Sun is high in the sky.

SEEING IMPOSSIBLE LIGHT

If the lens in a pair of Polaroid sunglasses only transmits vertically polarized light under normal circumstances, then if you take the glasses off and turn them through a right angle, so that the earpieces are at the top and bottom instead of on either side, the lenses will only transmit horizontally polarized light – in effect, you have stood the 'picket fence' on end. Horizontally polarized light cannot pass through a vertically oriented polarizer, so it is pretty obvious that if you take two pairs of Polaroid glasses and wear one while holding the other in front of one eye, so that the lens is rotated by a right angle in this way, you won't see anything at all through the two lined-up lenses. For once, the photons behave just as common sense predicts; try it and see (or rather, try it and *don't* see!). This is an example of a pair of 'crossed' polarizers.

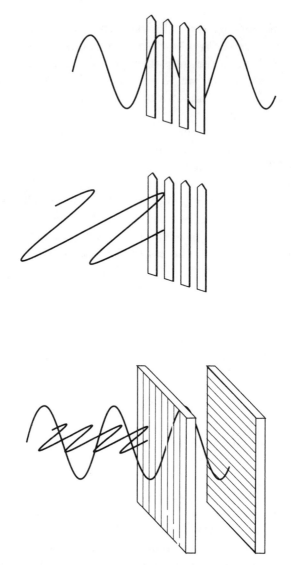

Figure 14
If light is a wave, it is easy to understand how 'vertically polarized' light can slip through a piece of polarizing material, represented here as a picket fence (top).
Obviously, horizontally polarized light will not be able to get through the pickets (middle). And two pieces of polarizing material that are 'crossed' at right angles will stop both vertically and horizontally polarized light (bottom).

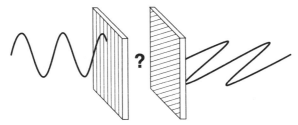

Figure 15
Strangely, if the second polarizer is at 45 degrees to the first, it doesn't block the vertically polarized light. Exactly half the light gets through, but now it is polarized at 45 degrees.

But the triumph of common sense is short-lived. In the everyday world, it seems equally obvious that if you have two lenses lined up in such a way that no light at all can get through, you still won't get any light going through if you put a third lens in between the two original lenses. This time, common sense is wrong. Take a third pair of sunglasses, and hold it so that it is oriented at 45 degrees to each of the first two pairs, with one lens lined up in between the pair of crossed lenses. Without the extra lens, no light is transmitted and the field of view is black; with the extra lens in place, some of the light gets through. Not as much as with only one lens in place (in fact, about a quarter as much), but some light definitely gets through three lenses in a row where it cannot get through two. Why?

You can begin to see what is going on by looking at the light passing through two polarizers which are not crossed at right angles but oriented at 45 degrees to each other. Forget the sunglasses, now; this is what you find in careful laboratory experiments using precise orientations and accurate measurements of the strength and orientation of the light beams involved. Let's say that the light going through the first polarizer emerges vertically polarized. What happens when it meets the second polarizer, in which the 'slots' are at 45 degrees to the vertical?

From the picket-fence analogy, you might expect that none of the light gets through. In fact, exactly half of the vertically polarized light passes through the second polarizer – and the light that emerges on the other side is now polarized at 45 degrees, lined up with the slots in the second polarizer. So when this reduced-strength light arrives at the third polarizer, crossed with the first one and therefore horizontal, the relative orientation between the light and the polarizer is again 45 degrees. Once again, half the light gets through – and now, the

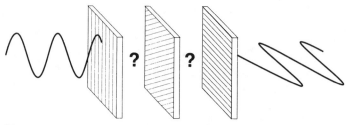

Figure 16
Stranger still, three polarizers successively twisted through 45 degrees allow a quarter of the light to pass, and it emerges horizontally polarized – even though no *light will get through if you take the middle polarizer away!*

emerging beam is horizontally polarized! By passing a vertically polarized light beam through two suitably oriented polarizers it has been weakened to a quarter (half of a half) of its original strength, and twisted so that it is now horizontally polarized.

The experiment can even be carried out with a beam of light so weak that we are dealing with individual photons passing through the crossed polarizers. Just as in the experiment with two holes, photons can be fired one at a time through the apparatus. When you do this, you find that what you expect to be a vertically polarized photon (that is, one that has passed through a vertically oriented polarizer) arriving at a polarizer inclined at 45 degrees has a 50:50 chance of being transmitted or being stopped. If 100 'vertically polarized' photons are sent through the second part of the experiment, 50 will be stopped and 50 will be transmitted but will now, if measured, turn out to be polarized at 45 degrees. At the next polarizer, horizontal compared with the original polarization of the photons, 25 out of the 50 survivors will be stopped, and 25 will be transmitted but will emerge with horizontal polarization.

Of course, the experiment can be carried out with the two polarizers at different relative orientations. If both are vertical, then all the vertically polarized photons get through. If the two polarizers are crossed at right angles, no photons get through. And for any in-between orientation there is a definite proportion of the photons that get through, ranging smoothly from 100 per cent down to zero. It seems as if each individual 'vertically polarized' photon actually has a well-defined probability of being oriented with a different polarization – zero chance of being horizontally polarized, but 50 per cent chance of being polarized at 45 degrees, a smaller chance of being oriented at, say, 30 degrees, and a greater chance of being oriented at 60 degrees. The photon itself is actually in an indeterminate condition, one of those 'superpositions of states', until the measurement of its polarization is made. Then it 'decides' whether or not it is polarized in the right way, and passes on or not according to the strict probability rules. As Paul Davies has put it:

> It must be emphasised that quantum indeterminacy does not merely mean that we cannot know which polarization direction the photon really possesses, it means that the notion of a photon with a definite polarization direction does not exist. There is an inherent uncertainty in the *identity* of the photon itself, not just in our knowledge of it.[1]

And there is more – much more.

A calcite crystal is different from Polaroid sunglasses in one important respect. When a beam of light interacts with the crystal, it doesn't just emerge as one polarized beam. Instead, the crystal splits the light into two beams, polarized at right angles to each other, that emerge at slightly different places on the other side of the crystal. Vertically polarized light seems to follow one path through the crystal, while horizontally polarized light follows a different path. By making sure that the light arriving at the crystal is polarized midway between these two orientations (by which we mean that the incident light has passed through a polarizer oriented at 45 degrees), we end up with two equally strong beams of light, each half the strength of the original beam, one vertically polarized and one horizontally polarized, moving parallel to one another.

When a single photon passes through the crystal, of course it has to 'decide' which path to follow, and it will emerge, experiments confirm, from one or the other channel, with the appropriate polarization.

If you actually do this experiment with a beam of light, you can also place another calcite crystal in the path of the two beams that emerge, in such a way that the vertically and horizontally polarized light beams are merged inside the crystal, back into a single beam, polarized at 45 degrees – the two crystals are the 'opposite way round' to each other in terms of their crystal structure and their effect on light.

But what happens to a single photon going through the crystals?

---

[1] Davies, *Other Worlds*, p. 121.

'Obviously', when it arrives at the first crystal it must still decide whether to be vertically or horizontally polarized, and which channel to follow.

The point is confirmed by one last refinement to the experiment. Suppose that we now block off one of the two routes, by sticking a piece of black material in between the two calcite crystals, where it will stop any light emerging from one channel but not the other. Suppose that this is set up to stop any horizontally polarized photons that emerge from the first crystal. This has been done in real experiments. Now only half of the photons arriving at the first crystal pass through the experiment and emerge on the other side of the second crystal – but they are all vertically polarized. Exactly the equivalent result occurs when we block off the vertical channel and let only horizontally polarized photons pass; another triumph for common sense.

But what happens when we remove the obstruction in the horizontal channel, and let photons pass, one at a time, through the whole experiment? Common sense says that they will all get through, and will emerge either vertically or horizontally polarized, with equal probability. Once it has decided which state it is in, an individual photon going into the second calcite crystal can hardly be expected to turn back into a 45-degree-polarized photon, can it? And yet, it does! When the beam is so weak that only single photons are passing through the experiment, the light behaves as if each individual photon has split in two and followed both paths through the experiment, recombining with itself to restore the original polarization. Every photon arriving at the first crystal passes right through the experiment and emerges on the other side, with its original polarization restored. The probability waves are seeking out every possible route from one side of the experiment to the other, and taking account of the entire set-up before 'deciding' how to behave, just as they use every part of the mirror in 'deciding' how to reflect. It seems as if each individual photon following either route through the experiment is aware whether or not the other channel has been blocked off, and has modified its behaviour accordingly. All of this is old hat, by the standards of quantum theory, and has been known for decades. But in the 1990s experimenters have come up with even more subtle experiments that show individual photons behaving as both particle and wave at the same time.

SHEDDING MORE LIGHT ON LIGHT

One of the key features of the standard interpretation of quantum mechanics – the Copenhagen Interpretation – is the idea, established by Niels Bohr, of complementarity. This says that where a quantum entity such as a photon has a dual nature, such as wave–particle duality, no experiment will ever show both facets of its character at once. You can, said Bohr, carry out experiments designed to measure wave properties of light, and, sure enough, you will measure wave properties. Or you can carry out experiments designed to detect photons as particles and, indeed, you will detect particles. But, he said, you will *never* see light acting as both wave and particle at the same time.

Well, he was wrong. In 1992 Japanese researchers, carrying out an experiment devised by an Indian team, did just that. They observed individual photons exhibiting wave-like properties and particle-like properties at the same time.

Just what this means for our understanding of the quantum world is, as yet, far from clear. The one certain thing is that it is bad news for the Copenhagen Interpretation in its standard form. But I don't find this too alarming, since it seems clear to me that the Copenhagen Interpretation is in any case far from being the best explanation of what quantum reality is all about. But the experiment is well worth looking at in a little detail, simply as an example of the weirdness of quantum reality.

One of the strangest aspects of the story is that before physicists could show photons behaving like waves they first had to establish, in the 1980s, that photons really exist. As I have mentioned, Albert Einstein introduced the idea of what are now called photons in his 1905 explanation of the photoelectric effect, and duly received the Nobel Prize for his pains. But from the early 1950s onwards several researchers, starting with David Bohm (of whom more later) realized that the photoelectric effect can be explained without invoking photons after all! Just by treating light as a varying electromagnetic field which interacts with a metal surface made up of individual atoms, which can only accept definite amounts of energy, it is still possible to explain the photoelectric effect. Planck himself would have been delighted at the news, and strictly speaking it means that Einstein did not deserve the Nobel Prize (at least, not for the work he was given it for); now, though, it is only a curiosity of the history of science, because, spurred on partly by these ideas, the experimenters really have proved that photons exist.

Making individual photons to study isn't quite as simple as taking a light with a dimmer switch and turning it down until it is so faint that photons are being emitted one at a time. The problem is partly that the light is being radiated from many different atoms, and partly that the individual atoms have some 'choice' over exactly which energy changes (transitions) are involved in emitting the light. The energy in the light has to come from somewhere, and it comes from electrons jumping down from one energy level to another inside the atoms, and losing energy in the process. In most cases, many transitions of this kind, over a range of energy levels, combine to produce the light. This brings in an averaging over probabilities rather like the averaging involved in Feynman's path integrals; it means that a very faint pulse of light may actually carry less energy than a single photon, because it represents an average over many quantum states (a superposition, like the dead-and-alive cat), most of which are empty and contain zero photons! These bizarre low-level pulses of light behave like waves, and can be made to interfere with one another in suitably subtle experiments.

To make a genuine single photon, a single atom has to be triggered into emitting just one pulse of energy in a single transition between well-defined energy levels. Then, there is no scope for any super-position and the photon emerges in a pure, single quantum state. The way the experimenters do this is to prepare calcium atoms that have been energized (or 'excited') with light from a laser. If you think of the electrons in an atom as sitting on the steps of an energy staircase, this is equivalent to taking one of the electrons and moving it up two steps higher than it belongs; it will sit trembling on the edge of the higher step for a moment, and then bounce back down, first onto the intermediate step just beneath its new position and then (after a pause lasting just 4.7 billionths of a second) back to where it belongs. Each downward bounce releases energy, in the form of a photon.

To catch a single photon, the excited calcium is monitored with a detector that responds to the first-level photon and opens a 'gate' that allows light to pass for a short time. The length of time the gate is open is matched to the length of time the atom stays in the intermediate excited state, so when the second photon is emitted it passes through the gate and into the experiment proper.

Alain Aspect and Philippe Grangier, working in Paris, were among the pioneers of this kind of research in the 1980s. Having obtained their photons, they sent them towards a kind of mirror called a beam-splitter, which allows half the light that falls on it to pass, and reflects the other half at right angles. This is rather like the way a calcite crystal splits a beam of light into two beams, although in this case no polarization effects are involved. It is easy to see how a wave can be divided into two in this way, and with light from conventional sources divided by a beam-splitter the beams can be recombined and allowed to interfere, confirming their wave nature (this has even been done with the very low-intensity conventional light sources that correspond to fractions of a photon). But if a particle arrives at the mirror, it must be either reflected or transmitted; it cannot do both.

When detectors are placed in the paths of the two beams from the splitter, and the experiment is carried out with the single photons from the excited calcium atoms, that is exactly what the researchers find. The photons always follow one path or the other; there is never a simultaneous 'click' of the detectors in both channels, which would imply that half the light had gone each way.

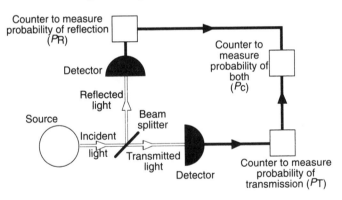

Figure 17
Can a single photon be split in half? If light really does come in the form of particles, then each photon arriving at the beam-splitter should either be reflected or refracted. According to quantum theory, the detectors should record perfect anticorrelation.

But this is not the end of the story. As Bohr would have predicted, when Aspect and Grangier looked for particles they found particles. What would happen if they looked for a wave, even though they 'knew' the calcium atoms were emitting photons?

To do this, they took the detectors out of the beams and replaced them by mirrors which recombined the light that had been divided by the beam-splitter. This is very much like a rerun of the experiment with two holes, and what they found was indeed that as more

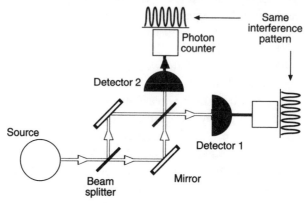

Figure 18
But when beams from an experiment like the one shown in Figure 17 are recombined in a second beam-splitter (which is 'running backwards'), they produce identical interference patterns, showing that even 'single photons' exhibit wave behaviour.

and more photons went through the experiment they built up the characteristic interference pattern corresponding to wave behaviour.

Using the same source of light, the Paris team could identify either particles or waves, in what seemed the perfect and definitive example of Bohr's complementarity at work. But hardly was the ink dry on the scientific papers announcing these results than three Indian scientists came up with a new suggestion for an experiment which could show *single* photons behaving *both* as particles *and* as waves *at the same time*.

The leading light in the Indian team was Dipankar Home, of the Bose Institute, in Calcutta. His colleagues were Partha Ghose, also of the Bose Institute, and Girish Agarwal, of the University of Hyderabad. The essential new step they introduced was to suggest replacing the beam-splitter mirror by another kind of beam-splitter, made of two right-angle triangular prisms almost touching one another.

The prisms are simple triangular blocks of transparent material, with one of the corners of the triangle making a right angle. With a single prism of this kind, when light comes in perpendicular to one of the 'square' faces, so that it hits the 'hypotenuse' face at 45 degrees on the inside, it is totally reflected, through a right angle, and comes out of the other 'square' face. If the two prisms are touching one another, hypotenuse to hypotenuse, to make a square block, light coming in perpendicular to one of the faces goes straight through the block and out the other side without being reflected at all. But if there is a tiny gap between the two hypotenuse faces, some of the light is reflected and some of it 'tunnels' across the gap and carries on in a straight line.

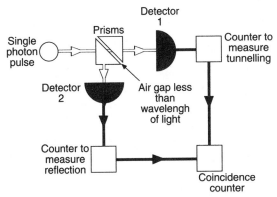

Figure 19
In one variation on the theme, experiments have been carried out with the beam-splitter replaced by two prisms separated by a tiny air gap. The light can only cross the gap by tunnelling, which involves acting as a wave. But the coincidence counter still records perfect anticoincidence, which is a particle property. The same *photons are caught behaving as particle and wave at the same time.*

The gap has to be really small for the trick to work – smaller than the wavelength of the light involved. In effect, if the gap is smaller than one wavelength some of the light can reach out across the gap without noticing it is there, and proceed on its way. As usual, probabilities and statistics come into the story. The smaller the gap, the bigger the proportion of the light that tunnels; so for a precisely set gap and a particular wavelength of light exactly half the light will be transmitted and half will be reflected. The essential point, though, is that only waves can tunnel in this way. Particles cannot tunnel.

This variation on the beam-splitter experiment was actually carried out, using pure single photons, by Yutaka Mizobuchi and Yoshiyuki Ohtake, of Hamamatsu Photonics in Hamakita. Some idea of the subtlety of their experiment can be gleaned from the fact that the size of the gap had to be controlled to within a few tens of a billionth of a metre, about one-tenth of the wavelength of the light involved. Once again, detectors were placed in the positions where the two beams of light, corresponding to reflection or transmission, should emerge from the prisms. Single photons cannot be split in half, so as usual there is a precise 50:50 chance that an individual photon arriving at the air gap will be reflected or transmitted. So if the two counters 'click' only in anticoincidence (that is, never at the same time as each other) that would be proof that light is travelling through the experiment in the form of photons.

But any photons that follow the straight-through path in the experiment can only have done so by tunnelling. In other words, by behaving as waves. When the experiment was carried out, the researchers did indeed find half the photons in each channel, confirming that they were behaving like waves at the gap and tunnelling. They also found that the detectors clicked in perfect anticoincidence, confirming that the photons were behaving as particles at the gap and not splitting in half. The experiment observes the same photons acting as both wave and particle at the same time (when confronted by the air gap), contradicting Bohr's basic tenet of complementarity. 'Three centuries after Newton,' says Home, 'we have to admit that we still cannot answer the question "what is light?"'; and he gleefully points to a remark made by Albert Einstein in 1951, in a letter to his old friend Michelangelo Besso: 'All these fifty years of conscious brooding have brought me no nearer to the answer to the question "what are light quanta?" Nowadays every Tom, Dick and Harry thinks he knows it, but he is mistaken.'[1] The point is reinforced by physicists in New Zealand, who have dreamed up an experiment (not carried out by 1994) which will, if quantum theory is correct, show a single photon in two different places at the same time.

SEEING DOUBLE

Of course, the photon isn't actually *in* two places at the same time. It just seems that way – another example of quantum non-locality, the 'spooky action at a distance' that so troubled Einstein.

The proposed experiment will use an array consisting of not just

[1] Both quotes from Dipankar Home and John Gribbin, 'What is light?', *New Scientist*, 2 November 1991.

one beam-splitting mirror, but three. After the original light beam is divided, the beams in both the two new channels will themselves be deflected onto beam-splitting mirrors, to produce a total of four possible routes for an individual photon through the experiment. Sensitive detectors in each of the four channels will record the arrival of any photon that threads its way through the apparatus to that particular spot.

With a beam of electromagnetic waves, there is no difficulty in predicting and understanding what will happen. The beam will be split in two at the first mirror, and each of the half-beams will be split in two at the corresponding mirror that it meets later. There will be four beams emerging from the experiment, each one-quarter as strong as the original, and all marching in step with one another.

So far so good; but this is not the experiment. All we have done so far is to set up a reference system of beams, which can be used to monitor the behaviour of individual photons sent in to the experiment from another source, by interfering with them. What Daniel Wallis and his colleagues at the University of Auckland propose is to take the set-up outlined above and shoot additional individual photons onto the first beam-splitting mirror. The photons will actually be injected at right angles to the reference beam of light, but this will not affect the behaviour of the mirror, which should send the photons, with equal probability, along one or other of the two channels, to the associated secondary mirrors.

Now, things start to get interesting. Assuming that *no* photons are being injected into the experiment, you might think that (after allowance is made for the reference beam) none of the four detectors on the other side of the experiment will record any photons. But you would be wrong. Just as an electron is allowed to produce a photon which it promptly reabsorbs, so nothing at all (the vacuum) is allowed to produce photons spontaneously, provided they quickly disappear back into the vacuum. This is an aspect of quantum uncertainty – to say that there was an absolutely zero probability of a photon being in a particular volume of space would imply absolute certainty, which is not allowed by the quantum rules. So there must be a small probability of a photon popping up anywhere at all. Anything which is not forbidden by the quantum rules seems to be obligatory, and, indeed, these quantum fluctuations of the vacuum, as they are known, are a well-established feature of the quantum world.

So, just occasionally, even when no photon is sent in to the experiment one of the detectors will click. Even more occasionally, two of the detectors will click together, because each of them has detected one of these 'virtual' photons popping into existence. When a single 'real' photon is fired into the experiment, it can only follow one path through the maze, and arrive at one detector, making one click – assuming that it behaves as a particle. So firing individual photons one after another into the experiment should increase the number of detections, with occasional coincidences caused by vacuum fluctuations affecting another detector at the same time as the 'real' photon arrives.

But it isn't that simple. Quantum theory says that there is actually an entanglement between the 'real' photons and the photons of the vacuum itself. In effect, there is interference. And, just as with the experiment with two holes, interference means that sometimes the two components add together and sometimes they cancel out. Depending on the properties of the incoming photons, in the experiment proposed by the New Zealand team there should sometimes be more coincidences, sometimes fewer, and sometimes the number of coincidences will stay the same. With the planned four-detector set-up, and photons fired in one at a time, the team expects to observe a large number of coincidences in one pair of detectors at the same time that the number of coincidences in the second pair of detectors stays at the level corresponding to simple vacuum fluctuations.

This would be a clear indication of quantum effects at work, and cannot be explained either in terms of pure classical waves or in terms of pure classical particles alone. The fact that an individual photon arriving at one detector *simultaneously* changes the probability of a virtual photon appearing out of nothing at all at a second detector on the other side of the experiment gives the illusion, when both detectors click together, that the single photon has arrived in two places at once. The original photon is detected in only one place, but its influence affects what is happening somewhere else at the same time.

It will be intriguing to see how the experiment turns out. It would be astonishing if the results turn out any different from the predictions of quantum theory; but among other things, if the results do turn out the way Wallis expects, they will reinforce the evidence that vacuum fluctuations really do occur. The idea of an active vacuum is an old one, but still worth a closer look.

SOMETHING FOR NOTHING

It isn't just photons that can appear out of nothing at all, as vacuum fluctuations. The quantum rules provide for a trade-off between uncertainty in energy and uncertainty in time. The energy needed to make a very light particle (such as a photon, which has zero rest mass, although it does carry energy) can pop up out of nothing at all for a relatively long time (only 'relatively' long; we are dealing in small fractions of a second here); but the energy needed to make more massive particles (such as an electron–positron pair) can only be 'borrowed' from the vacuum for a correspondingly shorter period of time (so the electron–positron pair quickly annihilate one another and give their borrowed energy back to the vacuum). 'Nothing at all' is actually best pictured as a seething maelstrom of activity in which all kinds of particles are flickering in and out of existence.

In the most extreme example of this, some cosmologists have seriously suggested that the entire Universe may be a quantum fluctuation. Since the Universe is about 15 billion years old and contains rather a lot of particles, this may seem hard to swallow at first. But it happens that the energy of a gravitational field is negative, in the same sense that mass energy is positive. If a tiny bubble of energy corresponding to the mass of the Universe popped into existence on the quantum scale, its mass energy and its gravitational energy could, the theory tells us, exactly balance one another, allowing the quantum universe to have essentially zero overall energy and therefore a very long lifetime. The final step in making the Universe out of nothing at all is to invoke a process called inflation, which swooshes this vastly subatomic seed up to about the size of a basketball in a fraction of a second, after which it continues expanding in a more sedate way.

But let's not worry too much at present about making universes out of nothing at all. Instead, I want to tell you about an experiment in which the activity of the vacuum can actually be detected, through its influence on atoms of sodium.

The right way to think of the vacuum is not as 'nothing at all' but as a superposition of many different states of the electromagnetic field (you can add in other fields as well, but let's keep it simple for now). The different states of the field are rather like the different notes that can be obtained from a single plucked guitar string, and (like the energy levels of an electron in an atom) they form an energy 'staircase' with the steps spaced at a distance corresponding to the energy of a single photon. When an atom emits a photon, what is happening is that the energy of the corresponding frequency of the vacuum field is increased by one unit, matching the decrease in energy of the electron in the atom. The temporary appearance of a virtual photon corresponds to the field energy moving up a step all by itself, and then falling back – like a guitar that plays random notes to itself, very softly.

Near an electrically conducting surface, however, the vacuum field and its fluctuations are modified, because the electric part of the field at the surface of a conductor must be zero. This eliminates some of the possible activity of the vacuum field, so that for an atom passing near the conductor there is more vacuum energy on the side of the atom further from the conductor. As a result, the atom ought to be pushed (or pulled, depending on your point of view) towards the conductor – there is a force attracting the atom to the plate.

This idea goes back to the 1940s, but the effect was only measured in 1993, by Ed Hinds and colleagues at Yale University. A related effect happens when two conducting plates are placed extremely close together in a vacuum; then, the modification of the vacuum field between the plates produces a force of attraction pulling them together. It is known as the Casimir effect, after the Dutch physicist Hendrik Casimir, and it has been measured many times using different sorts of conducting plates. But the Yale experiment is much more subtle, and more sensitive.

In this experiment, the researchers used two tiny glass plates coated with gold as their conductors. The two plates were wedged together to make a 'V' shape just a few millionths of a metre wide at the top, and atoms of sodium were sent through the gap of the 'V' at different heights. The actual separation between the plates at each height was measured to an accuracy of five billionths of a metre, using interference fringes produced by pure monochromatic light. So the experimenters knew exactly how close the atoms were passing to the plates, and could calculate how much they should be influenced by the vacuum force. As the atoms emerged on the other side of the 'V', they were monitored by bouncing laser beams off the atoms. The observed behaviour of the atoms exactly matched the predictions of quantum theory, allowing for the extra force, and did not match the predicted behaviour of atoms passing through a classical channel of the same width. 'Nothing at all' was *measured* to be having an effect on the individual sodium atoms.

I love the combination of the simplicity of the idea behind this experiment and the fine-scale subtlety of the way it was put into practice (*and* the fact that it shows quantum theory, once again, triumphant). It took more than 40 years from the time that theorists came up with the idea before experimenters could test it, but it was well worth the wait. It may take at least that long for some of the current crop of theoretical ideas to be tested, but those experiments, if they ever are carried out, will be even more spectacular. Would you believe, for example, that quantum theory tells us that teleportation – yes, 'Beam me up, Scotty' teleportation, just like in *Star Trek* – may actually be possible?

'BEAM ME ABOARD, SCOTTY'

Remember the EPR 'thought experiment', that was actually put into practice by Alain Aspect and his colleagues? They showed that a pair of photons which are produced in such a way that they must have opposite polarizations – but nobody knows what the polarizations are – remain in an entangled state as they fly in opposite directions across the Universe. When the polarization of one of the photons is measured, the other one, instantaneously, collapses into the opposite state. This entanglement and action at a distance is at the heart of the technique of quantum teleportation proposed by Charles Bennett, of the IBM Research Center at Yorktown Heights in New York, and published in the highly respectable journal *Physical Review Letters* in 1993. Apart from its science-fiction-like overtones, the important point about this work is that the team showed how to solve what seemed an insuperable quantum problem, using quantum techniques themselves.

In the classical, everyday world sending copies of things to distant places is routine. The obvious analogy with teleportation is the fax machine, which has the added advantage of leaving the original copy intact at its starting place while producing a duplicate at the destination. Newspapers and books[1] are reproduced in editions containing hundreds

---

[1] Well, *some* books, even if not this one!

of thousands of essentially identical copies, as far as their information content is concerned. But at the quantum level, copying runs into difficulties.

The first is simply a question of detail. The uncertainty principle makes it impossible to know every detail about every atom in, say, a sheet of paper, or even the exact position of every molecule of ink in the printing on the paper; so the faxed 'copy' can only ever be an approximation. In addition, scanning an object at the quantum level changes its quantum state – the very act of looking at something alters it, according to quantum theory. So even if you did obtain the information needed to build a copy of a quantum system, the original would be destroyed. In a way, this is actually more like the SF version of teleportation than the way a fax machine works. In SF, it is usually an essential feature of teleportation that the 'original' is destroyed – although several stories have looked at the unpleasant consequences of using teleportation devices to create multiple copies of human beings.

Classical information can be copied, but can only be transmitted at the speed of light (or less); quantum information cannot be copied ('a single quantum cannot be cloned', as the physicists quip), but sometimes, as in the EPR experiment, it seems to propagate instantly from one place to another. Bennett and his colleagues used a mixture of these classical and quantum features of a system to propose their teleportation device.

They describe this in terms of two people, Alice and Bob, who want to teleport an object. In this teleportation for beginners, the object to be teleported is simply a single particle – perhaps an electron – in a particular quantum state. At the beginning of the experiment, Alice and Bob are each given a box containing one member of a pair of entangled objects, equivalent to them each carrying one of the photons from the EPR experiment, without measuring its polarization. Then they go off on their travels across the Universe. Some time later – perhaps many years later – Alice wants to send another particle to Bob. All she has to do is to allow the 'new' particle to interact with her entangled particle, and to measure the outcome of their interaction. This both establishes and changes the state of her entangled particle, and instantaneously establishes and changes the state of Bob's entangled particle in an equivalent way.

Bob doesn't know this yet, because he is somewhere on the other side of the Universe. So now Alice has to send him a message, perhaps by radio, or perhaps by putting a notice in the newspaper that Bob reads every day, telling him the result of her measurement. This message contains only classical information, so she can send as many copies as she likes in as many newspapers or radio broadcasts as she likes. Eventually, Bob will get the message. Armed with the information about how the interaction between Alice's two particles turned out, Bob can now look at his own entangled particle and use the information to 'subtract out' the influence of his own original particle from its present state. What he is left with is an exact copy of the *other* particle – the one that Alice wanted to send to him. And she has done this without knowing where Bob is, or even speaking to him directly. The original version of the third particle was destroyed (changed into another quantum state) when Alice carried out her measurement on it, so Bob's version is unique, unlike a newspaper, and he is fully entitled to regard it as the original particle, conveyed to him by a combination of classical message and action at a distance.

This, Bennett stresses, defies no physical laws, and only permits teleportation to take place at less than the speed of light – Bob needs Alice's 'classical' message in order to untangle his particle properly, and if he looks at his particle too soon he will change its quantum state and destroy any prospect of untangling it in the right way. 'Alice's measurement forces the other EPR particle to change in such a way that the classical information that comes out of her measurement enables someone else to produce a perfect copy of what went in,' but 'it cannot take place instantaneously.'[1] It is, as one wag has remarked, 'teleportation, Jim, but not as we know it'. Given the ingenuity of the experimenters, though, there must be a good chance that before 40 more years have passed they will be sending electrons from one side of the lab to the other, or even around the world (if not across the Universe) in this way. It will be a neat trick, even if it has no practical implications. But there may even be practical implications, if not for this specific work then certainly for some related investigations of the mysteries of the quantum world. Bennett's fertile imagination does not stop at teleportation, and one of his other achievements, more obviously associated with the interests of IBM itself, concerns the possibility of using quantum mechanics to create an uncrackable code.

[1] Quotes from *Science News*, 10 April 1993.

QUANTUM CRYPTOGRAPHY

There is, of course, a connection with teleportation. The teleported particle contains information, and in principle this could be a message. A spy who was equipped with an entangled particle could use it to teleport another particle back to his superiors, and all the spy would have to do would be to send a straightforward communication in plain language reporting the outcome of the measurement made when the new particle (the 'message') interacted with the entangled particle. Anyone would be free to intercept the plain-language communication, which would be completely useless without the second entangled particle.

In fact, the investigation of ways to send uncrackable messages through quantum channels started before the teleportation work, and flourished in the 1980s. There are several approaches to the problem, but they all depend on code systems that use a 'key' of random numbers.

This kind of code is familiar from spy stories. The two people using the code are each equipped with an identical list of random numbers, a so-called 'pad', which can be as thick as a telephone book. The person sending the message turns it into numbers (perhaps as simply as by allotting the number 1 to letter A, 2 to B, and so on), and then chooses one of the pages of random numbers from the pad. The numbers on the pad are written out under the numbers corresponding to the letters in the message, and the pairs of numbers are added together. The coded message is then sent, along with information about which page on the pad was used, and at the other end the same set of random numbers are subtracted from the coded signal to restore the original message. The code is called the Vernam cipher, after the American Gilbert Vernam, who developed it during the First World War, and is sometimes referred to as the 'one-time-pad' technique, because spies were supplied with the random-number book in the form of a tear-off pad, so that each sheet could be used once and then destroyed (if the same set of random numbers from the same page of the pad are used to code more than one message, patterns recur which make it possible to crack the code).

This kind of code cannot be broken, unless the person who intercepts it also has a copy of the same one-time pad. The snag is that under the conditions in which espionage agents usually operate it is all too likely that the interested third party will get hold of a pad; worse, it is possible for the third party to have a copy of the pad, and to be breaking the code, without the two users of the code knowing.

Quantum physics offers a way around both problems. There is no need to keep the coded messages themselves secret, because, exactly like Alice's classical message to Bob, they are useless without the information that comes over a quantum channel, in this case, the key. What is needed is a way to communicate the key itself – a string of random numbers – from Alice to Bob in an uncrackable way. To make things as simple as possible, the string of numbers can be in binary arithmetic, a string of 0s and 1s like the code used by computers; so the key can be transmitted as any system of on/off, either/or signals.

Bennett and his colleagues have shown that you can do this using polarized light. The technique involves Alice sending Bob a stream of photons, polarized either up or across one of two agreed orientations (at 45 degrees to each other), but with the polarization of each photon chosen at random. Bob measures the polarization of the incoming photons, but for each measurement he can only align his detector with one of the agreed directions of polarization – again, chosen at random. In each case, he will get an 'answer' corresponding either to vertical (binary 1) or to horizontal (binary 0) polarization relative to his detector. He then tells Alice which orientations he used for each measurement, and she tells him which of these match the way the photons were sent (this communication can be by public telephone). Bob and Alice then throw out all the measurements for which Bob chose the 'wrong' polarization, and are left with a string of 1s and 0s, their secure key in binary code. It sounds like a tedious process when spelled out like this, but in the real world anyone using such a system would avoid the tedium by running it through a computer which did the donkey-work.

The great beauty of the technique is that the only way a third party can find out what code they are using is to 'eavesdrop' on the quantum communication channel, and measure the polarization of the photons as they pass through. But the act of measuring the polarization of a photon, as we saw earlier, changes the polarization! Even if the eavesdropper copies the measured photon and sends it on to Bob, it will have been randomized. Bob and Alice can check for this interference by standard techniques which, in effect, compare every fifth, or every seventh, or whatever, letter of the key, without revealing the whole key.

It may all sound far-fetched and improbable; but Bennett and his colleagues have actually built a system which works in this way. Admittedly, in the prototype the code messages are sent across a distance of only 30 cm; but that is because they built it on a desk top. In principle, polarized photons could be sent unaltered through several kilometres of optical fibre. And after all, when John Logie Baird built the first television transmitter it sent a picture over a distance of only a couple of metres.

The quantum cryptographers are already at work dreaming up better ways to pass on their keys. Artur Ekert, of the University of Oxford (who has also collaborated with Bennett), has shown how the required random string of digits could be obtained from a variation on the EPR experiment. The EPR photons are fired off in opposite directions, suitably entangled but as yet unmeasured, one beam to Alice and one to Bob. Alice and Bob can *each* measure the polarization of their photons, using detectors oriented at random along one of a set of previously agreed polarization directions. They then tell each other, over any ordinary public communications channel, which measurements they made, but *not* the results of those measurements. Finally, they discard the measurements where they used different orientations, and construct their secure key out of the results of the measurements where their polarization detectors were aligned the same way – making allowance, of course, for the fact that each pair of EPR photons has the opposite polarization, once they have been measured, so that Bob always gets a 1 where Alice gets a 0, and vice versa. And, once again, any attempt to 'tap' the quantum communication channel by looking at the photons before they get to Bob and Alice will scramble up their polarizations in a detectable way.

As these examples show, the quantum properties of photons themselves are now being used in practical ways – not yet commercially available, off-the-shelf quantum code machines or teleportation devices, but the beginnings of, literally, desk-top-sized prototypes. So the reality of photons, together with the fact that light behaves both as wave and particle, is beyond question. Just as the experimenters seem to be bringing even those bizarre quantum properties into the everyday world of engineering, though, other experiments have been upsetting the simple picture of the particle nature of the photon by probing 'within' it. Thanks to quantum uncertainty, even the inside of a photon must now, it seems, be thought of as a seething mass of particles. After all, a photon has more energy than the vacuum

– so if the vacuum can be full of virtual particles, why not a photon?

INSIDE THE PHOTON

The way I have described them so far, photons are simple entities that can only interact with other particles through the effects of the electromagnetic force. Since photons are 'made' of electromagnetism, how could it be otherwise? But apart from gravity (which is a very weak force and can be largely ignored for subatomic particles) and electromagnetism itself, there are two other forces which operate at the subatomic level. The weak nuclear force is associated with the behaviour of nuclei that results in radioactivity and nuclear decay, while the strong nuclear force holds the particles that make up atomic nuclei (protons and neutrons) together. In fact, protons and neutrons are themselves thought to be made up of more fundamental entities known as quarks, and the strong force operates between quarks. It all seems very neat and tidy. And yet, in some experiments involving energetic photons interacting with protons it seems as if the photons themselves are being influenced by the strong force – as if they can 'feel' the quarks inside the proton, not just the electric charge on the proton.

These tantalizing hints of some extra layer of activity involving photons encouraged scientists at the Desy laboratory, just outside Hamburg, to carry out very-high-energy experiments involving photons in the early 1990s. Those experiments showed that the photons do indeed behave as if they were complex entities, made up of an interior jumble of quarks, electrons and other particles. The explanation for this is exactly the same as the explanation of the quantum nature of the vacuum. The uncertainty in the amount of energy that a photon carries allows it to turn itself into a quark–antiquark pair (and other things) for a short time, just as the uncertainty about the zero point of energy in the vacuum allows electron–positron pairs (among other things) to pop in and out of existence. If the photon collides with a proton while it is in this state, the quarks 'inside' the photon will interact directly with the quarks inside the proton to produce a shower of other particles that can be detected using standard techniques.

These are new discoveries, whose implications are still being investigated and which will keep the experimenters busy for years to come. The essential feature of the discovery is, though, already clear. Having struggled to come to terms with the notion of wave–particle duality as a description of light, we now have to come to terms with the idea that light itself can change into matter and then back into light again, down at the level where a split second is measured in terms of the Planck time, 10^{-43} of a second.

Strange though this behaviour is, it does contribute to the pleasing symmetry between light and matter, waves and particles, that is such a strong feature of the quantum world. We have, after all, already met atoms that will, under the right circumstances, go 'two ways at once' through the equivalent of the experiment with two holes, interfering in the way that we normally expect light waves to behave; so we surely ought to be broad-minded enough to allow light 'waves' to behave not only as some special kind of particle (photons) but even, under the right circumstances, like the very particles of which atoms are ultimately composed.

But how *do* particles of matter, including atoms, behave? We have learned that in some sense they do not really exist, as particles, when nobody is looking at them – when no experiment is making a measurement of their position or other properties. Quantum entities exist as a superposition of states unless something from outside causes the probabilistic wave function to collapse. But what happens if we *keep* watching the particle, all the time? In this modern version of the kind of paradox made famous by the Greek philosopher Zeno of Elea, who lived in the fifth century BC, a watched atom can never change its quantum state, as long as it is being watched. Even if you prepare the atom in some unstable, excited high-energy state (like the atoms used to make the single photons involved in the paradoxical experiments described earlier), if you keep watching it the atom will stay in that state forever, trembling on the brink, but only able to jump down to a more stable lower-energy state when nobody is looking. The idea, which is a natural corollary to the idea that an unwatched quantum entity does *not* exist as a 'particle', had been around since the late 1970s. A watched quantum pot, theory says, never boils. And experiments made at the beginning of the 1990s bear this out.

WATCHING THE QUANTUM POT

Zeno demonstrated that everyday ideas about the nature of time and motion must be wrong, by presenting a series of paradoxes which 'prove' the impossible. In one example, an arrow is fired after a running deer. Because the arrow cannot be in two places at once, said Zeno, at every moment of time it must be at some definite place in the air between the archer and the deer. But if the arrow is at a single definite place, it is not moving. And if the arrow is not moving, it will never reach the deer.

When we are dealing with arrows and deer, there is no doubt that Zeno's conclusion is wrong. Of course, Zeno knew that. The question he highlighted with the aid of this 'paradox' was, *why* is it wrong? The puzzle can be resolved by using the mathematical techniques of calculus, which describe not only the position of the arrow at any moment, but also the way in which the position is changing at that instant. At another level, quantum ideas tell us that it is impossible to know the precise position and precise velocity of the arrow at any moment (indeed, they tell us that there is no such thing as a precise moment, since time itself is subject to uncertainty), blurring the edges of the argument and allowing the arrow to continue its flight. But the equivalent of Zeno's argument about the arrow really does apply to a 'pot' of a few thousand ions of beryllium.

An ion is simply an atom from which one or more electrons have been stripped off. This leaves the ion with an overall positive electric charge, which makes it possible to get hold of the ions with electric fields and keep them in one place in a kind of electric trap – the pot. Researchers at the US National Institute of Standards and Technology, in Boulder, Colorado, found a way to make the pot of beryllium ions boil, and to watch it while it was boiling – which stopped the boiling.

At the start of the experiment, the ions were all in the same quantum energy state, which the team called Level 1. By applying a burst of radio waves with a particular frequency to the ions for exactly 256 milliseconds, they could make all of the ions move up to a higher energy state, called Level 2. This was the equivalent of the pot boiling. But how and when do the ions actually make the transition from one quantum state to the other? Remember that they only ever decide which state they are in when the state is measured – when somebody takes a look at the ions.

Quantum theory tells us that the transition is not an all-or-nothing affair. The particular time interval in this experiment, 256 milliseconds, was chosen because for this particular system that is the characteristic time after which there is an almost exact 100 per cent probability that an individual ion will have made the transition to Level 2. Other quantum systems have different characteristic times (the half-life of radioactive atoms is a related concept), but the overall pattern of behaviour is the same. In this case, after 128 milliseconds (the 'half-life' of the transition[1]) there is an equal probability that an individual ion has made the transition and that it is still in Level 1. It is in a superposition of states. The probability gradually changes over the 256 milliseconds, from 100 per cent Level 1 to 100 per cent Level 2, and at any in-between time the ion is in an appropriate superposition of states, with the appropriate mixture of probabilities. But when it is observed, a quantum system must always be in one definite state or another; we can never 'see' a mixture of states.

If we could look at the ions halfway through the 256 milliseconds, theory says that they would be forced to choose between the two possible states, just as Schrödinger's cat has to 'decide' whether it is dead or alive when we look into its box. With equal probabilities, half the ions would go one way and half the other. Unlike the cat-in-the-box experiment, however, this theoretical prediction has actually been tested by experiment, just as Newton would have wished.

The NIST team developed a neat technique for looking at the ions while they were making up their minds about which state to be in. The team did this by shooting a very brief flicker of laser light into the quantum pot. The energy of the laser beam was matched to the energy of the ions in the pot, in such a way that it would leave ions in Level 2 unaffected, but would bounce ions in Level 1 up to a higher energy state, Level 3, from which they immediately (in much less than a millisecond) bounced back to Level 1. As they bounced back, these excited ions emitted characteristic photons, which could be detected and counted. The number of photons told the researchers how many ions were in Level 1 when the laser pulse hit them.

Sure enough, if the ions were 'looked at' by the laser pulse after 128 milliseconds, just half of them were found in Level 1. But if the

---

[1] The analogy with radioactive half-life is not exact, because in this case the transition is being 'pumped' from outside by the radio waves, which is why *all* the ions make the transition in just 256 milliseconds.

experimenters 'peeked' four times during the 256 milliseconds, at equal intervals, at the end of the experiment two-thirds of the ions were *still* in Level 1. And if they peeked 64 times (once every 4 milliseconds), almost all of the ions were still in Level 1. Even though the radio waves had been doing their best to warm the ions up, the watched quantum pot had refused to boil.

The reason is that after only 4 milliseconds the probability that an individual ion will have made the transition to Level 2 is only about 0.01 per cent. The probability wave associated with the ion has already spread out, but it is still mostly concentrated around the state corresponding to Level 1. So, naturally, the laser peeking at the ions finds that 99.99 per cent are still in Level 1. But it has done more than that. The act of looking at the ion has forced it to choose a quantum state, so it is now once again purely in Level 1. The quantum probability wave starts to spread out again, but after another 4 milliseconds another peek forces it to collapse back into the state corresponding to Level 1. The wave never gets a chance to spread far before another peek forces it back into Level 1, and at the end of the experiment the ions have had no opportunity to make the change to Level 2 without being observed.

In this experiment, there is still a tiny probability that an ion can make the transition in the 4-millisecond gap when it is not being observed, but only one ion in 10,000 will do so; the very close agreement between the results of the NIST experiment and the predictions of quantum theory shows, however, that if it were possible to monitor the ions all the time then none of them would ever change. If, as quantum theory suggests, the world only exists because it is being observed, then it is also true that the world only changes because it is not being observed all the time.

This casts an intriguing sidelight on the old philosophical question of whether or not a tree is really there when nobody is looking at it. One of the traditional arguments in favour of the continuing reality of the tree was that even when no human observer was looking at it, God was keeping watch; but on the latest evidence, in order for the tree to grow and change even God must blink, and rather rapidly!

So we can 'see' ions frozen into a fixed quantum state by watching them all the time. We can also, thanks to researchers at IBM's Almaden Research Center in San Jose, California, 'see' the waves of probability that determine the behaviour of electrons.

THE GREAT ELECTRONIC ROUND-UP

One of the neatest examples of the electron wave at work was developed in the 1990s by Franz Hasselbach and colleagues at the University of Tübingen, in Germany. They used a refinement of a device called an electron interferometer, which had itself been invented at Tübingen in the mid-1950s.

The electron interferometer is a version of the experiment with two holes. The electrons are sent in a beam towards a wire that has a negative electric charge. The negative charge in the wire repels the negative charge of the electrons, and the apparatus is designed with complete symmetry so that there is a 50:50 chance that an electron in the beam will pass by the wire on either side. Further along, there is a positively charged wire, so arranged that it will attract electrons passing by on either side and channel them back into a single path. Finally, a detector records the arrival of electrons on a screen, just as in the more familiar double-slit version of the experiment.

When electrons are sent through the interferometer one at a time, they build up an interference pattern on the screen at the other end, as if each individual electron had split in two when it passed the first wire and the two half-electrons had been recombined by the second wire and interfered with each other (I trust you are not surprised to learn this; by now, you ought to have been astonished only if I told you that electrons did *not* behave in this way). So far, just another version, albeit a particularly subtle one, of the experiment with two holes. But in 1992 the Tübingen team added a refinement.

In that set of experiments, they added a device called a Wien filter to the electron interferometer. The Wien filter is made up of a pair of electrically charged plates with a gap between them (in essence, a capacitor), with a magnetic field running across the gap at right angles. Charged particles, such as electrons, that move through the filter 'feel' both the electric and the magnetic fields; the two fields are balanced so that any charged particle passing through the filter will be slightly deflected, unless the particle is moving at a particular speed corresponding to the way the filter is set up. Now, the researchers destroyed the symmetry of the original experiment by setting up the Wien filter between the two charged wires of the electron interferometer in such a way that one half of the 'split' electron beam would feel a tug as it passed through, while the other half of the beam did not. As a result, half of the electron wave moved faster through the inter-ferometer, getting out of step with its counterpart. This changed the interference pattern produced on the screen, in exactly the way predicted by quantum theory – and it did so even when the electrons were passing through the interferometer one at a time. More proof that electrons behave as waves, but not quite the same thing as 'seeing' the waves themselves. That trick was achieved only in 1993, when the IBM researchers carried out the first round-up at the quantum corral.

As well as highlighting the reality of the quantum wave, this technique has practical implications, since it involves manipulating single atoms and arranging them on a surface – 'nanotechnology', which may soon lead to smaller, faster and more efficient computers, as well as other sub-microscopic artefacts which, some scientists believe, will transform society in a new industrial revolution. The IBM researchers used an instrument called a scanning tunnelling microscope (STM) to deposit 48 atoms of iron in a perfectly circular ring, just 14 billionths of a metre across,[1] on a flat copper surface. This was their 'quantum corral'. To an electron inside the ring of iron atoms, they present an impenetrable circular wall, and according to quantum theory electron waves inside the ring will reflect from the wall to form a standing wave – a pattern of ripples frozen in time like the string of a guitar endlessly playing the same note.

So, at least, says quantum theory. The density of electrons at any place inside the quantum corral can be measured, using the STM itself, and these measurements are converted into an image of what the pattern of electrons would look like if we had eyes that could see them directly. The image is exactly like that of a photograph of the ripples surrounding the point in a pond where a stone has been dropped in – the standing wave of the electrons themselves.

Electrons are seen to be behaving as waves. Even atoms, as we saw in the Prologue, behave as waves in variations on the theme of the experiment with two holes. And yet, it is worth pointing out that Hans Dehmelt, of the University of Washington, Seattle, won the Nobel Prize (in 1989) for his pioneering efforts, ultimately successful, to capture both individual electrons and individual atoms in magnetic 'boxes' (like the 'quantum pot' used in the Zeno's paradox experiment) and watch them behaving as particles. There is no way to 'see' a single

---

[1] A billionth of a metre is called a nanometre. The prefix 'nano' comes from the Greek *nanos*, meaning a dwarf, hence the term nanotechnology.

electron trapped in this way directly. But in the 1980s Dehmelt and his colleagues not only captured an individual atom of barium in one of these modified 'Penning traps' but actually photographed the atom by the natural blue light that it emitted. It appeared as a tiny blue dot in the centre of a vast sea of black in the photograph; if you are willing to accept that photography is as good as seeing with your own eyes (and after all, it is only through photography that we know what many of the distant galaxies and other objects in the Universe look like), it is now possible to see an individual atom.

Nevertheless, it is still possible for the philosophers and the quantum interpreters to debate whether or not the atom is there when nobody is photographing it. Perhaps I have given you enough examples of the strange truth of the quantum world, and it is time to get down to the nitty-gritty of explaining just what quantum reality really is all about, as I promised you I would. But before I become embroiled in the various interpretations of quantum reality on offer – most of them looking to an outsider like desperate remedies or counsels of despair – perhaps I should also give a clear picture of just what it is we are trying to explain, using two final examples of the strange behaviour of light itself.

WHEN IS A PHOTON?

One of the nice features of the way quantum physics has developed over the past few years is the way in which ideas that were dreamed up as 'thought experiments' and which were often never originally intended to be put into practice have, in fact, been turned into practical experiments that demonstrate vividly the strangeness of the quantum world. The archetypal example, of course, is the EPR experiment, adapted conceptually by John Bell and actually carried out by Alain Aspect's team. In that case, it took half a century for the thought experiment to become realized. But in other cases experimental progress has been much quicker.

John Wheeler, who was Richard Feynman's thesis supervisor, came up with a particularly neat proposal at the end of the 1970s, when working at the University of Austin, Texas. I touched on this 'delayed-choice' thought experiment in *In Search of Schrödinger's Cat*, little realizing that the experiment would actually be carried out within a couple of years of the book being published; I also mentioned there a literally cosmic version of the thought experiment, involving light from distant quasars. In the mid-1980s, *nobody* thought this version of the experiment could really be carried out; but in the mid-1990s there are real prospects that the measurements of quasar light that Wheeler discussed in thought-experiment terms less than 20 years ago will soon be realized.

The basic feature of the delayed-choice experiment is a variation on the experiment with two holes. We already know that if photons are fired through the experiment one at a time, they still build up an interference pattern on the screen on the far side of the experiment. It seems that each photon goes both ways through the experiment and interferes with itself. We also know that, if we set up a system to monitor which of the two slits the photons are going through, we always see individual photons passing through just one slit or the other – and in that case we do not get an interference pattern on the far screen. The behaviour of the photons at the slits is changed by how we choose to look at them.

Wheeler pointed out that it would be possible, in principle, to place the detectors that monitor the passage of the photons at an intermediate position, between the two slits and the far screen. We could look to see if the photons were behaving like particles or like waves *after* they had passed the slits, but *before* they arrived at the screen. Quantum theory says that by detecting a photon in one or the other channel we would collapse the wave function of the whole experiment so that no interference pattern would form; but by switching the detector off and choosing not to look at the photons as they flew past, we could restore the interference pattern. The way the light behaves at the slits is determined after it has passed the slits. Not only that, as Wheeler pointed out we do not actually have to decide whether or not to switch the detectors on until after the light has passed the slits – hence the name 'delayed-choice' experiment.

Like the story of Schrödinger's cat, this thought experiment highlights the absurdity of quantum mechanics. Unlike the experiment with Schrödinger's cat, it was carried out, in the mid-1980s, by two groups working independently of one another, one at the University of Maryland and one at the University of Munich. They actually used the variation on the theme where a single beam of laser light is split into two by a beam-splitting mirror. One of the split beams passes through a device called a phase-shifter, so that it marches slightly out of step (by a known amount) with the other half-beam, and then the two half-beams are recombined to make an interference pattern (this is exactly equivalent to the way electrons are 'split in half' and phase-shifted in the Tübingen experiment). Detectors known as Pockels cells could be placed in each of the split beams, to monitor the passage of photons, while detectors at the far end of the experiment looked at the recombined beam to see if it was producing interference or not.

The Pockels cells could be switched on and off very rapidly, within about 9 billionths of a second. But the length of the path of each beam from the beam-splitting mirror to the detectors was about 4.3 metres, and it would take a photon, travelling at the speed of light, 14.5 billionths of a second to travel that far. So the Pockels cells could be switched on (or off) *after* the light had passed the beam-splitter (and, of course, the decision on whether to switch the cells on or off had to be made by a computer, at random, with no human intervention). Both groups of researchers found results in agreement with quantum theory. With the detectors switched on, the light behaved like photons, with each photon travelling by one or other path, and no interference (there are, of course, a lot of photons in a burst of light 4.3 metres long, each of which had to 'decide' how to behave before it got to the detectors). With the detectors switched off, the light behaved like waves, even when a stream of individual photons was fired onto the beam-splitter mirror, with light seeming to follow both channels and definitely producing interference. The behaviour of the photons at the beam-splitter is changed by how we are *going to* look at them, *even when we have not yet made up our own minds about how we are going to look at them*!

This is an impressive example of a thought experiment becoming reality, but the apparent ability of the photons to predict in advance whether the detectors are going to be switched on or off when the photons arrive at them does only cover a span of a few billionths of a second, and you might not be too worried about such a modest amount of 'precognition'. This is where the cosmic version of Wheeler's thought experiment, which he dreamed up at the beginning of the 1980s, comes in.

Wheeler pointed out that a phenomenon known as gravitational lensing could produce a cosmic version of the experiment with two holes – in this case, the experiment with two paths. At the time, nobody was sure whether or not gravitational lensing of distant quasars would be detectable using telescopes on Earth, but since then several examples of the phenomenon have been discovered. What happens is that light from a quasar thousands of millions of light years away from us passes by an intervening galaxy somewhere along this long line of sight (a light year is the distance light can travel in one year; to put this in perspective, remember that it takes less than 500 seconds for light to reach us from the Sun, 150 million km away). If the galaxy and quasar are aligned in just the right way, the gravity of the galaxy will bend the light as it passes, so that photons from the quasar have a choice of two paths around the galaxy (rather like the way a single electron has a choice of paths around the charged wire in an electron interferometer), and will form two images of the quasar, one either side of the image of the galaxy, as viewed from Earth.

In principle, it would be possible to combine the light from the two images to make an interference pattern here on Earth. That would be 'proof' that the light was behaving like a wave, travelling both ways around the intervening galaxy. On the other hand, it would be possible to use Pockels cells (or something similar) to monitor the photons arriving at Earth from each of the two images of the quasar. In that case, quantum theory predicts, if we let the photons from the two images come together on a screen after they have been monitored by the Pockels cells, they will not make an interference pattern. That would be 'proof' that the light was behaving as particles, with each photon travelling one way or the other around the intervening galaxy.

The snag about turning this thought experiment into practical reality (the reason why, in 1980, nobody thought it could ever be anything but a thought experiment) was that although we can catch photons from either image of the quasar, the size of the galaxy that is bending the light is so great that it blurs out the information in the two beams. Any source of light has a characteristic 'coherence time' over which the light waves it emits are moving in step. Over longer timespans, the waves get out of step with one another in random and unpredictable ways. The difference in the lengths of the two paths for light around a galaxy is several light weeks, and this is much longer than the coherence time of light from a quasar. So the information in the light gets scrambled up, and cannot be used to make an interference pattern here on Earth.

In 1993, however, astronomers were excited by the news that another form of gravitational lensing had been discovered. This involves a dark, unseen massive object in our own Galaxy passing in front of a star in another galaxy and making the distant star 'twinkle' as it does so, as different gravitationally lensed images of the star pass across the field of view. These massive objects are probably no bigger than the planet Jupiter, and the path differences are correspondingly much smaller than those for galactic gravitational lensing. It should be possible, as the observations improve and telescopic techniques are refined, actually to detect interference patterns in the light from distant stars, or even quasars, in this way. And then it will be just a small step to put a Pockels cell into the experiment and stop the interference.

To put the significance of this in perspective, the photons arriving at the detectors strapped to our telescopes may have set out a billion years ago from a quasar 10^{22} km away from us. They have a 'choice' of two routes to Earth. They could go one way, they could go the other, or they could mysteriously split up and travel both ways at once. But which route they follow, starting out a billion years ago and 10^{22} km away, seems to depend on whether or not an astronomer on Earth, perhaps in the 1990s or the early 2000s, decides whether or not to switch on a Pockels cell attached to the telescope observing the photons.

The misconception in this image of what is going on, says Wheeler:

> is the assumption that a photon had some physical form before the astronomer observed it. Either it was a wave or a particle; either it went both ways around the galaxy or only one way. Actually, quantum phenomena are neither waves nor particles but are intrinsically undefined until the moment they are measured. In a sense, the British philosopher Bishop Berkeley was right when he asserted two centuries ago 'to be is to be perceived'.[1]

I'm not sure that this really helps. However you try to describe it, clearly something *very* strange is going on in the cosmic version of the delayed-choice experiment. The whole Universe seems to 'know', in advance, what experiment an individual human being is going to carry out, perhaps on a mountain top in Chile, some time in the next few years. Wheeler has actually gone so far as to suggest that the entire Universe only exists because someone is watching it – that everything, right back to the Big Bang some 15 billion years ago, remained undefined until it was noticed. This raises huge questions (like those raised by the cat-in-the-box thought experiment) about what kind of

---

[1] *Scientific American*, July 1992, p. 75.

creature qualifies as being alert enough to notice that it (and the rest of the Universe) exists, and collapses the cosmic wave function. These are among the issues that I want to get on to in the next chapter; but first, here is a rather sideways look at the whole business of collapsing wave functions, a thought experiment which says that the *lack* of an observation can make the wave function of a system collapse.

This wonderful example of the strangeness of the quantum world dates back to the early 1950s, and is known as 'Renninger's negative-result experiment', after the German physicist Mauritius Renninger who first thought of it. It is one of the easiest examples of quantum strangeness to understand – but not to explain.

In my slightly modified version of the thought experiment, imagine that we have a source which will emit a single quantum particle in a random direction (ordinary radioactive nuclei do exactly this, so there is nothing special about the source). This source is in the middle of a large hollow sphere, and the inner surface of the sphere is lined with material that will give a flash at the point where the particle hits it. The accepted quantum description of what happens when the source emits a particle is that a quantum probability wave spreads out evenly in all directions around the source, since there is an equal probability for the particle being emitted in any direction. When the probability wave reaches the inner surface of the spherical shell, there is just one flash of light as the wave collapses to a single point. The particle is only 'real' when it is being observed – when it makes the flash of light – not while it is travelling from the source to the sphere.

So far, simple enough. But now imagine that halfway between the source and the sphere there is a hemispherical shield, which blocks off exactly half of the outer sphere from the field of view of the source. Like the outer sphere, this inner hemispherical shell is lined with scintillating material that will flash when it is struck by a particle from the source. Now what happens when the source emits a particle?

It is possible to set up a very simple quantum description of the possible outcomes of this experiment, which involves only two final states. We are not interested in exactly where on the outer or inner spheres the particle makes a flash of light, only in which of the two spheres it strikes. Either the particle strikes the inner sphere and makes it flash, or it strikes the outer sphere and makes it flash. The way I have described the situation, there is an equal probability of either outcome of the experiment. Now, suppose that the source is once again triggered into emitting a particle. Once again, standard quantum theory describes this as an expanding spherical shell of probability, moving out evenly in all directions. We wait for a time longer than the time needed for it to reach the inner hemisphere, but too short for it to have reached the outer sphere, and see no flash on the inner sphere. So we *know* that the final state of the experiment will involve a flash on the outer sphere – the particle must have been emitted in the wrong direction to strike the inner hemisphere. From a 50:50 probability of the flash occurring either on the hemisphere or on the outer sphere, the quantum wave function has collapsed into a 100 per cent certainty that the flash will occur on the outer sphere. But this has happened without the observer actually 'observing' anything at all! It is purely a result of a change in the observer's *knowledge* about what is going on in the experiment. It requires an observer intelligent enough to infer what is happening, and what would have happened if the particle had been heading towards the inner hemisphere (so a cat, for example, clearly would not be intelligent enough to cause this particular collapse of a wave function). Under these circumstances, the *absence* of an observation can collapse the quantum wave function as effectively as an actual observation can. At least, so says the Copenhagen Interpretation.

This central role for the observer – not just any observer, but an intelligent observer – lies at the heart of the standard Copenhagen Interpretation of quantum mechanics, but is very hard to justify except as a desperate remedy to patch up the theory pragmatically into a form that could be used for 'cook book' quantum mechanicking, applying recipes to achieve certain ends without understanding what goes on while the quantum cake is baking.

Although most physicists have been happy for more than half a century to use the recipes and not worry about the quantum cookery, there have always been alternative ways of interpreting the strangeness of the quantum world. Unfortunately, although the debate has been lively the alternative interpretations on offer have all, until recently, been no less flawed than the Copenhagen Interpretation. But it is worth looking at these counsels of despair briefly, in order to see just how much any adequate interpretation of quantum theory will have to explain, preparing you to be suitably impressed when I unveil just such a superior theory at the end of this book.

CHAPTER FOUR

Desperate Remedies

One of the most remarkable features of quantum theory is that there are many different interpretations of what the theory 'really means', most of which are mutually contradictory as far as their philosophical basis goes, but all of which accurately explain the behaviour of known experiments and correctly predict the outcome of new experiments. They *all* pass Newton's test of a good theory! There is nothing like this in any other area of science – we do not, for example, have half a dozen or more different 'interpretations' of Einstein's general theory of relativity, the other great theory of twentieth-century physics.

The choice of interpretations of quantum theory is, in fact, reminiscent of the choice of paths a photon has through an experiment with two (or more) holes. The photon seems to be able to go both ways through the experiment at once, even though in the everyday world the two routes are mutually exclusive. Quantum theory seems to be open to many different mutually exclusive interpretations, but like the photon going both ways at once through the experiment with two holes all of them are, in some sense, correct. Rather than trying to say which single interpretation is 'the right one', some physicists (notably Heinz Pagels, author of *The Cosmic Code*) have argued that we should learn a little bit about the quantum world from each of the interpretations, considering all of them together in a kind of superposition of possibilities. Few of the experts, however, are broadminded enough to take this view. Instead, you tend to find that individual physicists (those who bother to think about these things at all, that is) cling stubbornly to the notion that their own favoured interpretation is correct, and that all of the other interpretations are 'obviously' wrong.

The nature of this debate – if that is not too gentle a word for such scientific mud-slinging – was deliciously brought out into the open in the mid-1980s, when Paul Davies (then a Professor of Physics at the

University of Newcastle upon Tyne) and Julian Brown (a BBC Radio producer) combined their talents to put together a BBC radio programme about quantum theory. They interviewed eight of the top quantum physicists of the time, asking them for their views on the quantum mysteries, and how those mysteries might be explained. After the broadcast, the full transcripts of the interviews, together with some introductory material, were published as a book, *The Ghost in the Atom*. There, the experts can each be seen, solemnly claiming that one particular interpretation is correct while the others are impossible. The only problem is, the experts do not agree on *which* interpretation is correct. Utterly sure of themselves, and with few exceptions, they all plump for different versions of reality and dismiss the others. The book brings out the cleavage not just between the different interpretations but between the interpreters themselves more clearly and accessibly than anything else I have seen, and I'll quote from it occasionally in this chapter to highlight those differences.

I don't intend, though, to give you an exhaustive review of every interpretation of quantum reality, but rather an overview of the handful or so of main contenders. My own view is that none of them provides a satisfactory explanation of how the world works, although, like Pagels, I agree that all of them provide useful insights. As I shall discuss in more detail in the next chapter, a theoretical 'model' of the world does not have to be perfect in order to be useful. And this is spelled out nowhere more clearly than by the example of the Copenhagen Interpretation, a model which is clearly flawed but which has provided the practical basis of quantum mechanicking for well over half a century.

THE COPENHAGEN COLLAPSE

The status of the Copenhagen Interpretation as the 'official' explanation of quantum reality is partly due to a historical accident, and partly due to a silly mistake made by one of the greatest mathematicians of this century. The historical accident is that it was the first interpretation that could be made to work, in the sense of providing recipes that quantum cooks who did not want to bother with the deeper mysteries and the philosophy could use to bake their quantum cakes (it also helped that it was put forward by a forceful personality, Niels Bohr, who seldom lost an argument). If the Copenhagen Interpretation worked at this practical level, few quantum mechanics bothered too much about its deeper implications.

Even in the mid-1980s, this official position still held sway, and not just among the less philosophical of the quantum cooks. Sir Rudolf Peierls, a physicist who was born in Berlin in 1907 and worked with many of the pioneers of quantum mechanics before settling in England, made this clear in his contribution to *The Ghost in the Atom*. 'I object to the term Copenhagen Interpretation,' he said, 'because this sounds as if there were several interpretations of quantum mechanics. There is only one. There is only one way in which you can understand quantum mechanics.'[1] There speaks a physicist of the old school, brought up in the tradition of Niels Bohr, Werner Heisenberg and Max Born.

By now, you should have a good idea of what the Copenhagen Interpretation is all about – the combination of complementarity, probability waves, and the collapse of the wave function – and I do not need to elaborate the details again. But remember that one leg of the tripod, Bohr's understanding of complementarity, has now been called into doubt by those experiments which show a single photon behaving both as a wave and as a particle in the same experiment. Remember also that to the Copenhagen school of thought a quantum entity, such as an electron or a photon, *does not have* properties, such as position and momentum, except when these properties are being measured. It is not just that we do not know what the values of these properties are; the theory says that those properties do not exist unless they are observed.

This brings into sharp focus the great problem with the Copenhagen Interpretation. *When* (or *where*) does the collapse of the wave function take place? Can a Geiger counter detect the emission of a radioactive particle from an atom and collapse the wave function of the whole system involved in Schrödinger's cat-in-a-box experiment? Seemingly not, especially in the light of the kind of experiment thought up by Renninger, in which the *absence* of a measurement is responsible for the collapse of the wave function! So is consciousness, even intelligence, an essential component in the collapse of wave functions?

Philosophically inclined physicists have argued about just where to

---

[1] Davies and Brown, *The Ghost in the Atom*, p. 71.

put the interface between the everyday world and the quantum world ever since the Copenhagen Interpretation was first put forward. Strict Copenhagenists hold that what we regard as physical attributes of an electron, say, are nothing more (or less) than relationships between electrons and measuring devices, and that these properties 'belong' to the whole system, not to the electrons. In a talk to the British Association for the Advancement of Science, at Keele University in August 1993, the American physicist David Mermin tossed out a particularly apt analogy for this explanation of what is going on.

Psychologists and biologists argue fiercely about the nature of intelligence, the extent to which it is inherited and how much it is a result of environmental influences and education. They have developed so-called 'IQ tests', which measure something called the 'intelligence quotient' of human beings. But although many years ago many people believed that IQ tests provided a measure of intelligence, it is now generally accepted that what IQ tests measure is the ability of people to do IQ tests. Innate intelligence may be a factor in determining this ability, but it is not the only factor. The result of the 'experiment' (getting someone to take an IQ test) depends on the nature of the experiment itself (to take a trivial example, if the test is written in Russian, and you don't understand Russian, there is no way you will get a high score on the test).

In a similar way, if we set out to measure the momentum, say, of an electron, what we are actually measuring is the ability of an electron to answer questions about momentum. The electron may, indeed, not have any such property as momentum, in the way we think of it in the everyday world, but it has other attributes which cause it to respond to questions about momentum in a certain way. We get experimental results – 'answers' – which we interpret as measures of momentum. But they are only telling us about the ability of electrons to respond to momentum tests, not their real momentum, just as the results of IQ measurements only tell us about the ability of people to respond to IQ tests, not their real intelligence.

Nick Herbert, an American physicist, has another analogy. Bohr said that isolated material particles do not exist, but are abstractions which we only identify through their interactions with other systems – as, for example, when we 'measure' the 'momentum' of an electron. This, says Herbert, is like a rainbow.[1] A rainbow does not exist as a

---

[1] Herbert, *Quantum Reality*, p. 162.

material object, and it appears in a different place to each observer. No two people ever see the same rainbow (indeed, each of your two eyes 'sees' a slightly different rainbow). But it is 'real' – it can be photographed. Equally, though, it is not real *unless* it is being observed, or photographed. In the same way, according to Bohr, the properties of a quantum entity such as an electron are a kind of illusion brought about by the interaction of the quantum entity with the experimental arrangement.

In this basic version of the Copenhagen Interpretation, the 'fact' is the record of the result of a measurement – the click of a Geiger counter, or the appearance of a flash of light marking the arrival of an electron at a detector screen. But since even the measuring devices are themselves made of electrons and atoms and other quantum entities, how can they avoid being described in the same terms as other quantum entities? The Geiger counter itself is, in principle, described by a quantum probability wave; it exists in a superposition of states ('click' and 'no click') until it is measured. We can imagine the detector itself being 'made real' by being monitored by a second detector, but then the second detector (like Schrödinger's cat) exists in a super-position of states until it is monitored by a third detector, and so on in an infinite regress. This is what leads some quantum interpreters to conclude that the something special that causes the collapse of wave functions goes on inside the brains of intelligent observers.

I THINK, THEREFORE

This is still the Copenhagen Interpretation, or at least a long-standing variation of it. According to Peierls (who, as we have seen, is an unreconstructed Copenhagenist), 'the moment at which you can throw away one possibility and keep only the other is when you finally become *conscious* of the fact that the experiment has given one result.'[1]

This is the line of reasoning that leads John Wheeler to infer that the Universe only exists because we are looking at it. The quantum mechanical description is seen, in this interpretation, in terms of *knowledge*, and the existence of the mind is absolutely crucial. This is

---

[1] Davies and Brown, *The Ghost in the Atom*, p. 73.

quite different from the suggestion that size has something to do with the distinction between the quantum world and the everyday world, although that idea also comes under the umbrella of the Copenhagen Interpretation. The problem with that idea is, again, where do you draw the line? In his book *The Emperor's New Mind* Roger Penrose, of Oxford University, argues (unconvincingly, in my view) that gravity has something to do with it. Gravity is a very weak force, and can be entirely ignored for entities such as electrons. Perhaps, if you think along these lines, when enough matter is present for gravity to be noticeable it destroys the 'quantumness' of an object and turns it into an everyday, 'classical' object. Penrose develops much more sophisticated arguments involving the way information is lost in black holes and how this might be compensated for by quantum activity elsewhere in the Universe; but the whole package is deeply unconvincing. Slightly more plausibly, David Bohm has suggested that heat may be responsible for blurring the edges of the quantum world. Thinking along these lines, every atom and every electron is constantly being jiggled about by random thermal motion, and perhaps this is what destroys the quantumness of entities once they reach a certain size and contain enough particles jostling against one another.

But the quantum interpreters who see it all as being 'in the mind' will have none of this. They will tell you that even an object as large as the Moon, full of atoms held together by gravity and jiggling about with the random thermal motion appropriate to its temperature, does not exist when nobody is looking at it. David Mermin, of Cornell University, is one of the physicists who have argued along these lines. The Moon doesn't simply disappear when nobody is looking at it, they say; rather, like the beryllium ions in the quantum pot described in Chapter Three, if nobody is looking at the Moon all its atoms and electrons and other quantum components start to become uncertain about their quantum states. The probability waves spread out, very slowly, from the states they were in when they were last observed; the whole Moon starts to dissolve away into a quantum ghost. But because the Moon is so big, the process is very slow. It doesn't take a few nanoseconds, but millions (perhaps billions) of years for the Moon to dissolve away into quantum uncertainty. And long before that happens somebody looks at it and collapses it back into a nice, well-defined state, with an accurately located centre of mass in a particular orbit around the Earth. The apparent existence of the Moon (and everything else) as a real object is then explained as simply another example of the quantum pot-watching effect, on this interpretation.

John Bell summed up the situation succinctly, referring to what happens when electrons are fired at a scintillator screen, the screen is photographed, and somebody looks at the photograph to find out the result of the experiment:

> There is a romantic alternative [to the idea that the division between the quantum world and the everyday world is a matter of size]. It accepts that there is a division, whether sharp or smooth ... between 'quantum' and 'classical'. But instead of putting this division somewhere between small and big, it puts it between 'matter' (so to speak) and 'mind'. When we try to complete as far as possible the quantum theoretic account of the electron gun, we include first the scintillation screen, and then the photographic film, and then the developing chemicals, and then the eye of the experimenter ... and then (why not?) her brain. For the brain is made of atoms, of electrons and nuclei, and so why should we hesitate to apply wave mechanics ... at least if we were smart enough to do the calculations for such a complicated assembly of atoms? But beyond the brain is ... the mind. Surely the mind is not material? Surely here at last we come to something which is distinctly different from the glass screen, and the gelatine film.[1]

Quantum interpreters who try to develop these ideas sometimes suggest that the brain itself is in some sense a special kind of quantum system that operates in a holistic, or non-linear way, peculiarly suited to collapsing wave functions. Certainly, quantum processes are involved in thinking and consciousness, as Henry Stapp, of the University of California, Berkeley, has spelled out. Human nerves, including those in the brain, operate by transmitting electrical impulses, and also by transmitting impulses across synapses (which you can think of as junctions between nerves), chemically. A pulse coming along a nerve will trigger the release of calcium ions, which travel across the gap and trigger the next pulse of activity. A typical calcium ion involved in this process will travel about 50 billionths of a metre in a time of 200 millionths of a second. 'Simple estimates of the uncertainty principle,' says Stapp, 'show that the wave packet of the calcium ion must grow to a size many orders larger than the size of the calcium ion itself. Hence the idea of a single classical trajectory becomes

---

[1] Bell, *Speakable and Unspeakable in Quantum Mechanics*, p. 191. I should emphasize that this is not Bell presenting his own view of quantum reality, but Bell summarizing the arguments of people like Eugene Wigner and John Wheeler.

inappropriate: quantum concepts must in principle be used.'[1]

This is true enough, and even obvious, once it is pointed out; but it is no different, qualitatively, from Bell's comment that the brain is made of atoms, and must therefore obey the laws of wave mechanics. It does not mean, for example, that these quantum properties of the human brain imply that no artificial computer brain can ever be conscious, although some people have tried to argue this. After all, electronic computers are also made of atoms, and obey the laws of quantum mechanics; if it turned out that the specific properties of calcium ions spreading out into regions of quantum uncertainty at synapses were an essential component of consciousness, for example, then it would be straightforward (in principle) to construct artificial computer brains that incorporated this kind of behaviour.

But enough is enough. Although some people have gone further up these mystic paths, there is no need. I have shown you where the Copenhagen Interpretation leads you, if you let it, and have (I hope) persuaded you that it is not entirely a satisfactory explanation of quantum reality. Its success, as I said, was in large measure due to the historical accident of being the first fully worked-out interpretation, and of being championed by a strong personality. The Nobel prize-winning physicist Murray Gell-Mann commented as long ago as 1976 that 'Niels Bohr brain-washed a whole generation of physicists into believing that the problem had been solved';[2] but one reason for Bohr's success with his brainwashing was that the only rival interpretation at that time seemed to have been undermined by a calculation carried out by the mathematician John von Neumann. The truth is, though, that von Neumann had erred.

VON NEUMANN'S SILLY MISTAKE

Von Neumann's mistake was particularly unfortunate, because the interpretation of quantum mechanics that it seemed to rule out actually resembles our naive view of reality much more closely than the Copenhagen Interpretation does. Physicists (like most scientists) are

---

[1] Stapp, *Mind, Matter, and Quantum Mechanics*, p. 152.
[2] Gell-Mann, *The Nature of the Physical Universe* (New York: Wiley), p. 29.

remarkably conservative, and tend to cling on to old ideas for as long as they possibly can, until irrefutable experimental evidence forces them to be abandoned; given this pattern of behaviour, it seems highly likely that in a straight fight with the Copenhagen Interpretation the alternative 'pilot-wave' version of what are known as 'hidden-variables' theories would have won hands down. A generation of physicists would have been brought up thinking that hidden-variables theories were the standard way to explain quantum reality, and the Copenhagen Interpretation would be remembered as a curious alternative put forward by Niels Bohr, if not quite in his dotage then certainly after his best work was behind him.

The point about hidden-variables theories is that an entity such as an electron may exist as a real particle, in the everyday sense of the term, with a real momentum and a real position at all times, but we cannot measure its properties with unlimited accuracy. The behaviour of particles in the quantum world is, on this picture, determined by some extra phenomenon – usually described in terms of a new field – which varies in a way that cannot be directly observed. The hidden variations of the new field determine the behaviour of particles at the quantum level, and if physicists knew what the hidden variables were they could use them to predict the actual outcomes of measurements, not just the probabilities of different outcomes. They could, for example, calculate whether Schrödinger's cat was alive or dead without opening the box to take a look.

The standard hidden-variables theory was put forward in its original form by Louis de Broglie in 1925. He had been born in 1892 (and lived until 1987), but had made a late start on his scientific career, partly because his education was interrupted by the First World War. He had been the first person to realize that electrons could be described in terms of waves, and in the mid-1920s he was trying to reconcile this discovery with the fact that electrons could also be described as particles. He came close to finding a fruitful way forward for the quantum interpreters. But unfortunately, although a member of the French aristocracy (he inherited both the French title *Duc* and the German title *Prinz* in 1960, on the death of his elder brother) de Broglie was not such a forceful character as Bohr, and didn't fight very hard for his idea when it was questioned in the 1930s. The essence of the idea is that an electron, for example, is a 'real' particle (in the everyday sense of the term), but that its behaviour is determined by the way it is pushed around by the so-called pilot wave, which obeys the rules of quantum probability.

The idea, which had never found favour with the Copenhagen school, ran into seemingly terminal difficulties when von Neumann published a seminal book on quantum theory in 1932. Among other things, that book included what seemed to be a mathematical proof that no hidden-variables theory could ever properly describe the behaviour of entities in the quantum world.

Physicists took this at face value, because von Neumann was one of the top mathematicians of his day. He had been born in 1903, in Budapest (his given name was originally Johann, but he used 'John' later in his life). In 1928 he invented the branch of mathematics known as games theory, which involves making mathematical models (sets of equations) to determine the best strategy for playing a game – how to achieve the most wins or, at least, how to avoid losses. This became a major branch of mathematics because of its application to war 'games' and economic models. He was also the first person to suggest that a conscious observer might be needed to 'collapse the wave function' and select one quantum alternative out of a superposition of states.

Von Neumann moved to the United States in 1930, and in 1933 he became the youngest member of the Institute of Advanced Studies at Princeton, which had just been established (partly to provide a base for Albert Einstein). He was involved in the pioneering development of electronic computers (in some circles, computers are still referred to as 'von Neumann machines'), and the development of the atomic and hydrogen bombs. Although he died young, in 1957, he had made a major impact on the science of the twentieth century. Von Neumann was nobody's fool; but even geniuses can sometimes slip up.

The slip involved, literally, the way things add up. In mathematics, if it doesn't matter in which order a certain operation is carried out, then the equations involved are said to commute. For example, 3 + 2 is the same as 2 + 3. Addition is a commutative property. But if the order in which the operation is carried out does matter, then the property does not commute. For example, 3 − 2 is *not* the same as 2 − 3. Subtraction is a non-commutative property. In the quantum world, it happens that even addition is not always a simply commutative property. In general, the order in which things happen affects the final outcome of a series of interactions. This is rather like cooking – when you are baking a cake, you will get distinctly different results if you carry out the operations 'add half a pint of water and bake for 30 minutes' or 'bake for 30 minutes and add half a pint of water'.

I won't go into details, but in von Neumann's 'proof' that hidden-variables theories cannot work he used the fact that a particular property of a quantum system obeys the commutative rules *on average* and applied this rule to individual components of the quantum system. This is a bit like saying that if the average height of every member of a class of schoolchildren is 1.2 metres, then the height of each child in the class is 1.2 metres. That is certainly one way to make the average right, but it is not the only way (nor, indeed, the most likely way). It would be silly to assume that every child had the average height.

It does take a little more mathematical insight than just taking averages to uncover the flaw in von Neumann's argument, but it still ought to be obvious to a competent mathematician. One such mathematician, Grete Hermann, pointed out the flaw in 1935, but was ignored. Everybody else continued to believe in von Neumann's proof until 1966, when John Bell showed that it was based on this false assumption. Two decades later, Bell expressed his surprise at what he had discovered:

> The von Neumann proof, if you actually come to grips with it, falls apart in your hands! There is *nothing* to it. It's not just flawed, it's *silly*! ... When you translate [his assumptions] into terms of physical disposition, they're nonsense. You may quote me on that: The proof of von Neumann is not merely false but *foolish*![1]

Writing in 1993, David Mermin referred to the generations of graduate students who might have been tempted to try to construct hidden-variables theories but had been 'beaten into submission' by the claim that von Neumann had proved that it could not be done. He said that von Neumann's 'no-hidden-variables proof' was based on an assumption so silly 'that one is led to wonder whether the proof was ever studied by either the students or those who appealed to it to rescue them from speculative adventures' in the realms of quantum interpretation.[2]

I have laboured the point a little for two reasons. First, because it shows that physicists can be just as gullible as anyone else in accepting an idea because 'everybody knows' that it is true, and because it is

---

[1] Interview in *Omni*, May 1988, p. 88.
[2] *Reviews of Modern Physics* 65 (1993), p. 803.

written in a famous book, without bothering to check the facts for themselves. Secondly, because of this pervasive influence of von Neumann's 'proof', many popular and semi-popular accounts of quantum theory, as well as some textbooks, still say that hidden-variables theories are impossible, even though Bell showed that the proof was flawed back in 1966. Don't believe them. Hidden-variables theories (or interpretations) *can* be made to work, with one important proviso that I shall come on to shortly. The surprising thing is that one person, at least, was not afraid to attempt to construct such a theory in the 1950s, and was not beaten into submission by being hit over the head (metaphorically speaking) with von Neumann's 'proof'. His name was David Bohm, and over the years he developed, with some help, a hidden-variables interpretation of quantum mechanics which works just as well as the Copenhagen Interpretation, but gives a completely different view of quantum reality.

THE UNDIVIDED WHOLE

Bohm summed up his own view of the nature of reality, sharply contrasting with that of Peierls, in *The Ghost in the Atom*. Asked whether he thought that the external world exists independently of our observations, he replied, 'Every physicist really believes that,' and went on to say that 'the universe as a whole does not depend on us ... I don't think [the mind] has a significant effect on atoms.'[1]

It's probably significant that Bohm was a member of a later generation of physicists than the quantum pioneers. Born in 1917, he only began his systematic construction of an alternative quantum interpretation in the early 1950s, 20 years after the Copenhagen Interpretation had taken up its position at the centre of the stage. It may also be significant that he was born in America, and brought up literally an ocean away from the powerful influence of Niels Bohr.

I've had a soft spot for Bohm ever since I discovered that he was first turned on to science by reading science fiction at the age of eight, and then discovered astronomy books – almost exactly the way I became interested in science, at about the same age, 30 years after he

[1] Davies and Brown, *The Ghost in the Atom*, pp. 119–20.

did. During the Second World War, Bohm was a graduate student working with Robert Oppenheimer in California, and made a small contribution to the Manhattan Project. Afterwards, he went on to Princeton University, and wrote a book trying to explain quantum theory from what he understood to be Niels Bohr's point of view. It was this attempt at explaining the standard interpretation that made him realize that he didn't understand what Bohr was getting at, and led him to develop his own interpretation.

About the same time that he began to develop these heretical views, Bohm's personal life was thrown into turmoil. He was called before the Un-American Activities Committee of the House of Representatives, and asked to testify about the politics of some of the scientists he had known at Berkeley, while working on the Manhattan Project. This was in the early days of the Cold War, at the end of the 1940s, when the US Administration became paranoid about the possibility of communist infiltrators giving away atomic secrets to the Soviet Union. Bohm refused, on principle, to answer questions about the personal lives of his colleagues, using the Fifth Amendment, which gives the right not to testify if the testimony may incriminate you.

At the time, the story made a minor splash, and was then forgotten. But the communist witch-hunt then began to get into its stride. Two years later, Bohm was indicted for contempt of Congress, and brought to trial; although he was acquitted, the mud slung at him during the trial, at the beginning of the McCarthy era, stuck, and he found it impossible to get a job in the United States. He moved to Europe, and settled at Birkbeck College, in London, where his quantum interpretation was developed over the next four decades.

As his response to the Un-American Activities Committee shows, Bohm was not one to be intimidated by authority or to toe the party line (ironically, in view of the implicit accusation made against him). That von Neumann claimed to have proved that hidden-variables theories were impossible didn't stop Bohm investigating that approach; he didn't find the flaw in von Neumann's argument, but by constructing a workable hidden-variables theory he showed that there must be one. Either that, or Bohm's theory was wrong. He died in 1992, just when alternatives to the Copenhagen Interpretation were at last beginning to be taken seriously by more than just a handful of physicists, but having had the satisfaction of seeing Bell establish which one of the two was wrong (Bell's discovery of von Neumann's mistake doesn't prove that Bohm's theory is *right*, of course, but it removes a major obstacle from its path).

Bohm's interpretation of quantum uncertainty was that particles always have a distinct position and velocity, but that any attempt to measure these properties will destroy information about them by altering the pilot wave associated with the particles. Poking the pilot wave in one place (perhaps by measuring the position of an electron) will immediately alter the shape of the pilot wave everywhere, affecting all of the particles under its influence.

This contains two key concepts. First, because it is the shape of the pilot wave that determines how it influences particles, it doesn't matter how strong (or how weak) the wave is at any location. As long as the wave is there, changing its shape will affect the particles. Secondly, the pilot wave responds instantaneously, everywhere, to a disturbance at one locality. The wave itself is non-local.

This is the one proviso that I referred to earlier. In 1966, Bell proved that hidden-variables theories could be made to work *provided that you accept non-locality*. The Aspect experiment is a specific example of non-locality at work – measuring the polarization state of one photon immediately determines the polarization state of the other photon, even if it is on the other side of the Universe.

But surely, you may say, I described the Aspect experiment in terms of the Copenhagen Interpretation? I did indeed. If Bell had found that *only* hidden variables theories require us to accept non-locality, that would have been a powerful reason for abandoning this kind of quantum interpretation. But he didn't. What he found is that *any* interpretation of quantum reality must involve non-locality!

Strictly speaking, this is a slight simplification. Bell found that if his famous inequality were violated that would imply abandoning the concept of 'local reality'. 'Local', in this connection, means no communication faster than the speed of light; 'reality' means that the world exists independently of our observations of it. By showing that nature violates the Bell inequality, the Aspect experiment (and other experiments since) has shown that one of the two has to go. This is an even more dramatic conclusion than you might realize at first, since the Bell inequality does not, in fact, depend upon quantum mechanics at all. If the Bell inequality is violated (which it is), then local reality must be abandoned *even if quantum mechanics is completely wrong*. The result of the Aspect experiment shows that the Universe is not 'local and real', *whatever* kind of scientific description you might dream up to describe how it works. If you want to believe there is a real world out there, you *cannot* do without non-locality; if you want to believe that no form of communication takes place faster than the speed of light, you *cannot* have a real world, independent of the observer.

Bell, who was born in 1928 and died in 1990, was even more distant in time than Bohm from the great days of the quantum pioneers, and never could understand why people were so ready to take the Copenhagen Interpretation as gospel. The de Broglie/Bohm idea of particle *and* wave working together 'seems to me so natural and simple, to resolve the wave–particle dilemma in such a clear and ordinary way, that it is a great mystery to me that it was so generally ignored', he said.[1] And he wasn't at all bothered by the notion of influences travelling faster than light, even if that meant (as it does) backwards in time. He would, he said, rather give up Einstein's special theory of relativity, if necessary going back to the idea of an ether (or, at least, to a preferred frame of reference) than give up the notion of reality:

> One wants to be able to take a realistic view of the world, to talk about the world as if it is really there, even when it is not being observed. I certainly believe in a world that was here before me, and will be here after me, and I believe that you are part of it! And I believe that most physicists take this point of view when they are being pushed into a corner by philosophers.[2]

Bohm developed further the idea that everything is connected to everything else, and affected (instantaneously) by everything that happens to everything else, through the pilot wave. Seemingly independent entities, going about their business with no apparent connection, are in fact, he suggested, each responding to some underlying process at work. A very simplistic analogy might be with the shadow of a dancer, thrown onto two screens on opposite sides of a stage by spotlights. As the dancer moves about the stage, each of the shadows changes. If you could see only the shadows, it would seem as if they were interacting with one another in some mysterious way, involving action at a distance; in fact they are both responding to a deeper underlying reality. In later developments of his ideas, Bohm proposed that the basic underlying order of the world consists of a field made up of an infinite number of overlapping waves, and that the overlapping

---

[1] Bell, *Speakable and Unspeakable*, p. 191.
[2] Davies and Brown, *The Ghost in the Atom*, p. 50.

of the waves produces local effects which we perceive as particles.

All of these ideas, and especially the notion of a pilot wave which is aware of conditions everywhere in the Universe and guides its particle(s) accordingly, are strongly reminiscent of Richard Feynman's sum-over-histories approach to quantum mechanics. Instead of saying that 'the photon' travels by every possible route to a mirror and then up to your eyes to make a reflected image, we can say that 'the pilot wave' travels by every possible route, and then 'tells' the photon which path to follow. Just a year younger than Bohm, Feynman too, of course, was distanced from the Copenhagen interpreters by both time and space, and came up with new ideas decades after that interpretation was established. Somehow, though, until very recently Feynman's ideas seem to have been regarded as more reputable than those of Bohm ('more' reputable, but not really fully respectable; even now the sum-over-histories approach to solving quantum problems is regarded as a bit weird by many physicists, even if it does work). Both these ideas, however, are conceptually related to another very strange way of interpreting the nature of quantum reality, which involves not just non-locality, or photons following all possible trajectories, but an infinite array of *universes*, fulfilling every possible outcome of every possible quantum choice, and doing so (although this is not always acknowledged by advocates of this interpretation) in a decidedly non-local way.

A PROLIFERATION OF UNIVERSES

For obvious reasons (they will soon become obvious, if they are not already), this is known as the 'many-worlds' interpretation of quantum mechanics. It has long been a favourite of mine, partly because I was never impressed by the Copenhagen Interpretation and this seemed the best alternative, and partly because it offers such wonderful opportunities for writing science fiction stories. But the story of the many-worlds interpretation has become more complicated, as it has become more popular and has, as a result, split, amoeba-like, into three different many-worlds theories. At the same time, an even better interpretation, which I discuss in the Epilogue, has appeared, to charm anyone dissatisfied with the options on offer for the past 40 years. I am no longer *quite* so enthusiastic about the many-worlds interpretation as I used to be; but it is still at least as good as the Copenhagen Interpretation, and it still provides fertile ground for science fiction stories; so here it is, in all its glory.

The basic idea of the many-worlds theory is that every time the Universe is faced with a choice at the quantum level, the entire Universe splits into as many copies of itself as it takes to carry out every possible option. A simple way to picture this is with the aid of Schrödinger's venerable cat-in-a-box. In that quantum thought experiment, there are only two choices. Either the radioactive atom decays, and the cat is killed, or it doesn't decay, and the cat lives. The conventional, Copenhagen Interpretation, remember, says that *neither* option is 'real' until an intelligent observer opens the box to take a look. Until that happens, everything inside the box exists in a super-position of states, so that the cat is somehow neither dead nor alive until it is observed. The many-worlds interpretation says that *both* options become real, immediately that the system is faced with the choice, but that the Universe divides in two. In one copy of the Universe, the experimenter opens the box and finds a live cat, while in the other version of reality the experimenter opens the box and finds a dead cat. The crucial point, though, is that on this interpretation the cat was *really* either alive or dead inside the box before the experimenter looked; there was no mysterious superposition of states, and no collapse of a wave function at the moment of observation. Each observer thinks that he or she is inhabiting a unique universe, and there is no way for people in the two universes to communicate with one another.

The many-worlds interpretation was developed by Hugh Everett in 1957, when he was a student working under the supervision of John Wheeler. At the time, Wheeler endorsed the idea – but some indication of his relative enthusiasm for the notion (compared with his enthusiasm for, say, the Wheeler–Feynman absorber theory) might, with hindsight, be apparent from the fact that, even though he was the supervisor and Everett the student, the many-worlds interpretation is sometimes referred to as the 'Everett–Wheeler' theory, but never as the 'Wheeler–Everett' theory. A few years later, Wheeler changed his mind about the many-worlds interpretation, deciding that although it makes exactly the same predictions as the Copenhagen Interpretation in every con-ceivable experimental test, it carries too much 'metaphysical baggage' to be taken seriously. This objection is a matter of taste; the whole business of a superposition of states and collapsing wave functions carries its own load of metaphysical baggage, and some people (myself included) find that package harder to swallow than the many-worlds idea. But Wheeler does have a point.

The problem is that in its original form the many-worlds interpretation requires an infinite number of universes, each splitting into infinitely more versions of reality every split second, as all the atoms and particles in the universe(s) are faced with quantum choices and follow every possible route into the future at once. The usual way to think of this splitting of universes is in human terms – that there might be a 'parallel world' in which the South won the American Civil War, one in which the communists never seized power in Russia, and so on. As I said, delightful stuff for the science fiction writers, and seemingly reasonable enough, at this human level. Everybody enjoys speculating 'what if' some key event in history had turned out differently. But is it so reasonable if we have to allow for every single tiny quantum choice to turn out in every possible way? And if that isn't reasonable, but the big choices affecting human history are the cause of a proliferation of universes, then we are back to the problem of deciding where you draw the line between the quantum world and the everyday world, and puzzling over whether or not the implications of a quantum choice have to be big enough for an intelligent observer to notice before it has any effect.

In spite of these difficulties, some cosmologists have taken up the many-worlds interpretation, and pulled it out of the obscurity it languished in for almost 30 years after Everett came up with the idea. The reason for their enthusiasm is that one great advantage of the many-worlds theory is that it does away with the need for either an intelligent observer or a measuring device 'outside the system' to collapse wave functions and make reality real. We are back with the puzzle of 'Wigner's friend' – if Wigner's friend looks in the box to see if the cat is dead or alive, but doesn't tell anyone, then the friend *also* exists in a superposition of states, until Wigner asks him what has happened. Then Wigner exists in a superposition of states until another observer checks the outcome of the experiment with him, and so on, literally forever. So what made the *Universe* real, and not a superposition of states?

Wheeler has tried to argue that our observations (or those of any intelligent observer) *now* can somehow feed back into the past and collapse the wave function of the Universe all the way back to the Big

Bang (and this from a man who has argued that the many-worlds interpretation carries too great a metaphysical burden to be taken seriously!). But since we are part of the system (the system being the entire Universe in this case), this is a doubtful argument. For the Universe to exist as one reality, not a superposition of states, the Copenhagen Interpretation strictly speaking requires the existence of an observer *outside* the Universe to do the collapsing of the wave functions. So some cosmologists have turned to the many-worlds interpretation, preferring to argue that there really are many universes, each occupying its own region of space and time, and all tracing their origins back to the Big Bang. The resulting mathematical description of the Universe(s) is complex, but some successes with this approach have been claimed. For example, researchers such as Stephen Hawking suggest that although there may be an infinite variety of universes in some sense 'alongside' each other, the most common kind of universe, and therefore the one in which we are most likely to find ourselves, should look very much like the Universe we actually do live in.

The physicist who most strongly supports the many-worlds idea in the 1990s is probably David Deutsch, of the University of Oxford. He describes it as 'the simplest interpretation of quantum theory',[1] and he uses a version of the many-worlds interpretation to provide a different explanation of what is going on in the experiment with two holes.

If you carry out the experiment with single photons in the usual way and get an interference pattern, it seems natural to interpret this as telling us that *something* has gone through both holes of the experiment. Various interpretations describe this something in terms of the probability wave, or the pilot wave, or the photon itself mysteriously existing in two (or more) places at once, even though whenever we look to see which slit it is going through we always find the whole photon at just one slit (and, of course, the interference pattern then disappears). Deutsch points out that the interference is produced exactly as if ghost photons are passing through the 'other' slit (provided we are not looking) and interfering with our real photon(s) to make the interference pattern. He then says that these extra photons are not 'ghosts' at all, but are real photons following the

---

[1] Davies and Brown, *The Ghost in the Atom*, p. 84. In the same interview, Paul Davies comments that the many-worlds interpretation is 'cheap on assumptions but expensive on universes'.

alternative quantum paths in the universe(s) next door.

In the experiment with two holes, according to Deutsch, when a photon is faced with a choice of two slits to go through, the Universe divides in two, and in one version of reality the photon goes one way, while in the other it goes the other way. But then we bring the two possible paths the photon could have followed back together again, so that they interfere and produce the interference pattern. In doing so, according to Deutsch – and this is a neat development beyond Everett's original idea – we have made the two versions of reality fuse back together again; they only existed as separate realities during the time that the photon was travelling through the experiment. The fact that we do observe interference, even when single photons are fired one at a time through the experiment with two holes, is seen by Deutsch as proof that all possible quantum variations on the universal theme really do exist 'side by side' in some fashion. Expressed like this, the many-worlds interpretation also looks like a variation on Feynman's 'sum-over-histories' approach. But how 'real' are the alternate histories?

Deutsch has dreamed up an experiment which, he says, will be able to tell us whether or not other universes really exist. It isn't possible to carry out this experiment yet, but it might be possible within a few decades – certainly within a human lifetime – if computer technology keeps on developing at its present rate.

His proposal is to build a computer 'brain' which has a direct awareness of what is going on at the quantum level. This superbrain is then set the task of observing a quantum system in which there are just two equally probable possible outcomes of a measurement – perhaps a measurement of the polarization of a photon, which must be oriented in one of two possible ways according to the way the experiment has been set up. If Deutsch's version of the many-worlds theory is correct, the superbrain will split into two copies of itself, and each copy will record one of the two possible results of the measurement. But instead of making a note of exactly what it observes, the superbrain will simply make a record of the fact that it is observing one, and only one, of the two possible outcomes.

The brain writes down exactly the same thing in both the parallel realities, testifying that it is observing only one reality. Then the two realities are brought back together in some interference process (perhaps by scrambling up the polarization states of the photon again). 'If the conventional interpretation is true,' says Deutsch, 'then at some time during [the superbrain's] deliberations all the universes but one will have disappeared,' as the wave functions collapse during the quantum measurement, and there will be no interference. But if the many-worlds interpretation is true, then there will still be interference even though the brain remembers having observed only one intermediate reality. But the brain does not remember *either* of the two intermediate possibilities as such; it simply remembers that it observed only one intermediate quantum state! If it had written down *which* intermediate state it was observing (exactly equivalent to monitoring which hole the photon passes through), that reality would have become set, and could not have been fused back with its counterpart to create the interference. 'It is a necessary consequence of the other things [the superbrain] does, that he must wipe out the memory of which one of those two possibilities he observed.'[1] The outcome of the experiment – interference – involves both intermediate states coexisting, but the memory is of being in 'a' single state. So the Universe must have split in two.

In spite of its appealing simplicity (in terms of assumptions, if not number of universes), there are still difficulties facing any version of many-worlds theory. The most striking is that it is non-local with a vengeance – if we carry out the experiment with two holes and allow the interference pattern to form, then according to Deutsch's variation on the theme you might see the splitting and rejoining as a fairly local phenomenon, going on in one corner of our laboratory, and of no significance to the rest of the Universe. But if we look to see which hole the photon goes through, we prevent the interference pattern forming, which means that the Universe has split itself into two copies, one for each possible route of the photon. Perhaps the choice of which hole the photon went through is of no great consequence to the Universe at large; but still, in principle, this split changes the quantum state of the entire Universe(s), instantaneously.

This doesn't seem to worry Deutsch, partly because he has a rather different view of time from our everyday notion of something flowing from the past, into the present and on to the future. In his book *The Fabric of Reality* he has argued that there is no 'flow' of time, and that there is no single present moment, except subjectively. If time really does 'flow', he says (echoing the arguments put forward by J. W. Dunne in the 1930s), then there must be a second sort of time which

---

[1] Quotes in this paragraph from Davies and Brown, *The Ghost in the Atom*, pp. 99 and 100.

is used to measure the way 'now' moves forward from one moment to the next, and a third sort of time with which to measure that time, and so on. There may be differences between the past and future – a series of snapshots of a human being as a baby, child and adult can be put in the right order easily enough – but that doesn't necessarily mean that anything is actually moving from the past to the future. In a dramatic leap of imagination, Deutsch suggests that there is no fundamental difference between snapshots of other times and snapshots of other universes; that the 'past' and 'future' are just special cases of Everett's many worlds.

This is drawing us into deep waters, upon which I do not wish to venture now. Since I am not convinced that Deutsch's variation on the many-worlds theme is the best way to understand quantum reality in the first place, there doesn't seem to be much point in going into great detail about what these ideas imply for our understanding of time.

One reason why I am not convinced by Deutsch's arguments is that he still seems to leave room for measurements and observations (and intelligence) to play a special part in what happens to reality. If the 'superbrain' experiment produces interference when the brain writes down only that it is seeing one reality, without specifying which one, but produces no interference when the brain writes down which reality it is observing, we are right back with the puzzle of photons that go 'both ways' through the experiment if we are not looking, but only one way when we do look. Personally, I'd be happier with the naive version of Everett's theory, in which the Universe is constantly splitting into multiple versions of reality, which can never communicate with one another. But there are yet more variations on this basic theme which ought to be mentioned before I move on to something else.

VARIATIONS ON A QUANTUM THEME

Many-worlds theory has been one of the growth areas in quantum interpretation since I wrote *In Search of Schrödinger's Cat*, largely because of the cosmological problems I have already mentioned. In the middle of the 1990s, the buzz about these ideas mainly concerned two related variations on the theme, which go by the names 'many-minds' and 'many-histories' interpretations.

Some idea of the interest in the possibilities opened up by the many-worlds interpretation can be gleaned from a quick run-down of some of the researchers involved. I have already mentioned David Deutsch, in Oxford; others include Dieter Zeh and Ernst Joos (University of Heidelberg), Claus Keifer (Zurich Institute for Theoretical Physics), Jonathan Halliwell (MIT), Wojciech Zurek (Los Alamos National Laboratory), Thanu Padmanabhan (Tata Institute, Bombay), Murray Gell-Mann (CalTech), James Hartle (University of California, Santa Barbara), David Albert (Columbia University) and Barry Loewer (Rutgers University). When the journal *Physics Today* published an article by Zurek on one aspect of this work, in October 1991, it provoked so many letters in response that just the ones *Physics Today* chose to publish, plus Zurek's response, took up eight pages of the magazine. There is a *lot* of interest in these ideas in the world of physics in the 1990s!

That particular article by Zurek also described another aspect of some of these approaches to quantum reality, a phenomenon dubbed 'decoherence'. This has to do with the amount of information we actually have about a quantum system, and the amount of information that would be needed to specify completely the quantum state of that system.

Take an electron, for example. The state of an electron associated with a hydrogen atom can be completely specified by just three numbers, corresponding to three 'degrees of freedom' (I am ignoring the spin of the electron to keep this simple). This is rather like the way in which the position of a balloon floating in a room can be specified by just three numbers, its perpendicular distances from two adjoining walls and the floor. You need more parameters to specify more complicated systems, because they have more degrees of freedom, and generally speaking you need three times as many numbers as there are particles in the system to specify its quantum state.

Padmanabhan has used the traditional example of a cat to hammer the point home.[1] A cat weighing a kilogram, he points out, might contain about 10^{26} atoms, so even ignoring the question of what individual electrons are doing, we need three times that many numbers to specify the quantum state of the cat. Our description of the cat

[1] *New Scientist*, 10 October 1992.

simply does not operate at this level, and when we say 'there is a cat sitting in the corner of the room' there are many quantum states which would match that description.

The effect of ignoring many degrees of freedom, proponents of this idea argue, is to make the object – the cat, in this case – behave like a 'classical' object, instead of like a quantum object. It is by ignoring degrees of freedom that we 'make' objects behave classically. This, according to supporters of this interpretation, is true even in the experiment with two holes. When we look to see the photon going through one hole, we are ignoring the existence of the other hole, and making the system behave classically; when we allow the photon to 'see' both holes, we are using all the information available to describe the experiment, and it behaves quantum mechanically.

> A system appears to behave more and more classically as we start ignoring large numbers of internal parameters of the system. This theory suggests that if we could devise an experiment to measure all the parameters specifying a cat, we would find that the cat behaves as quantum mechanically as an electron, and could exist in a combined state in which it is both dead and alive.[1]

It is our ignorance that makes things behave classically, and the extent of our ignorance is bigger for bigger objects that are composed of more quantum entities. Which naturally suggests to some researchers that this 'decoherence' is a good way to account for the fact that the Universe at large seems to behave like a classical system.

This is where the 'many histories' come into the story. Zurek has drawn an analogy between the way the Universe got to be the way it is and what happens to a collection of stable atoms and unstable radioactive atoms. As time passes, the unstable atoms will decay, and be transformed by nuclear processes into longer-lived atoms. So, whatever mixture we start out with, the end product is a collection of stable atoms. Quantum mechanics allows us to consider the possibility of many different quantum states of the Universe 'emerging' from the Big Bang. 'Only certain stable states,' Zurek says, 'will be left on the scene.' What decides which variations survive is how well they are correlated, in effect, with themselves – histories which tell a consistent story survive better than histories which are inherently unpredictable. These are precisely the histories which are most closely represented

---

[1] Padmanabhan, *New Scientist*, 10 October 1992.

by classical descriptions. Zurek calls this a 'predictability sieve', and says that 'the pure states selected by the predictability sieve turn out to be the familiar coherent states'.[1]

In Padmanabhan's terms, the Universe behaves as though it is classical because there are a lot of other universes that we are ignorant about – hence, 'many histories'. And, once again, strong echoes of Feynman's sum-over-histories approach. The new ingredient, though, is the requirement that the history we *perceive* should be consistent. The correlation between our memories and the records of past events is a central concept in Zurek's interpretation, and what we perceive, in this picture, is not the wave function of the entire Universe, but a few characteristic features of a branch – or even a bundle of branches – that is consistent with all of the events that are included in the description of the world by the observer. Observers may remember things, and agree with other observers on what 'the' history of the Universe is, even though there are other histories going on, unknown to that particular set of observers.

Late in 1993, quantum physicists proposed a feasible experiment which will reveal whether history really exists, or is simply a consistent set of present memories. This is the temporal equivalent of the Bell inequalities. Some physicists have argued that Bell's description of separated events occurring at a single moment in time can be turned around to describe events that follow one another in time but occur at the same place – in the same quantum system. Juan Paz, of the Los Alamos National Laboratory, and Günter Mahler, of the Santa Fe Institute (both in New Mexico) have shown how this can be turned into a practicable experiment to determine whether history really does exist in the way that common sense tells us it does.

The proposed experiments involve a sequence of controlled measurements on identically prepared systems. Ideal candidates would be beryllium ions, which have a well-defined set of electron energy levels and have already been used in similar quantum experiments – the 'quantum Zeno effect' mentioned in Chapter Three. But this time the electrons associated with the beryllium ions would be made to jump about between *four* different energy levels.

The ions would be prepared by using laser light to drive continuous electron oscillations between two of the chosen energy levels, and then to stimulate jumps from each of these levels to one of two higher

---

[1] *Physics Today*, April 1993.

energy levels. The 'temporal Bell inequality' predicts that the number of electrons ending up in different energy levels will depend in a certain way on the order in which the various possible transitions are stimulated.

This is a practicable experiment. Paz and Mahler have shown how measurements of the final state of the system can reveal how it arrived in that state. Common sense would say that there must have been a continuous history, in which the electrons went through several states in a well-defined sequence to get from their initial state to their final state. Just as Bell set up his equations in such a way that they conform to common sense (so that *violation* of the Bell inequality proves that there really is 'spooky action at a distance'), so the equivalent equations describing this experiment are set up to match common sense. If the outcome of the experiment agrees with the 'temporal Bell inequality', then common sense rules the quantum world. But if it turns out that the temporal Bell inequality is violated, that will show that there were no well-defined 'in-between' states – that, as Paz and Mahler put it, 'between actual measurement events (set at will by the initial preparation and the final reading of the state) histories are not an element of reality'.[1]

By analogy with the spatial version of the Bell test, if the inequality is violated then quantum events would be correlated across time (between initial and final states) without passing through any in-between states (without having a 'trajectory' through time). The Aspect experiment shows quantum entities behaving as if the space between them did not exist; the new experiment will (unless everything we have learned about the quantum world is wrong) show quantum entities behaving as if the time between them did not exist.

It should be no surprise, by now, that quantum physicists *expect* the inequality to be violated when the experiment is carried out. Indeed, since the experiment is so similar, in practical terms, to the quantum pot-watching experiment, it may well have been carried out by the time you read these words; I am confident that the outcome will be as quantum physics suggests, and not in line with common sense.

This is not quite as alarming as it sounds, because it is essentially a property of a pure quantum system. Where very many quantum particles are involved in a system (such as a human being, or a cat)

[1] *Physical Review Letters* 71 (1993), p. 3235.

the 'quantumness' may become blurred, if the idea of 'decoherence' is correct. So according to Paz and Mahler 'violations of temporal Bell inequalities can be made to disappear by increasing the strength of the interaction with the environment',[1] and history may be real for historians, even if it is not real for an electron.

But, as always seems to be the case in quantum physics, there are other interpretations, and one school of thought holds that although historians (and the rest of us) may 'remember' a coherent history, this does not necessarily mean that there really *was* a unique single history. The alternative 'many-minds' idea comes into the story in large measure from the work of David Albert. According to his idea, when an intelligent being interacts with a quantum system the brain of the intelligent being itself acquires a complexity determined by the complexity of the quantum system. Like Deutsch's hypothetical superbrain, it splits into as many states as it takes to 'see' every possible quantum alternative, but each split consciousness is only aware of observing one outcome to the experiment. If you actually did the cat-in-the-box experiment, according to Albert, when you opened the lid of the box you would actually see both outcomes of the experiment, and both would be noted as 'real' by your brain. But the two aspects of your mind could never communicate their feelings and beliefs about the outcome of the experiment to each other.

I have great difficulty taking any of this seriously. First of all, it puts the nature of consciousness and intelligence right back in the middle of the quantum debate. Secondly, it seems to pull the rug from under one of the central features of the quantum world, the probabilistic nature of the outcome of experiments. If *every* possibility is experienced as real by one of my minds, what does it mean to talk about the probability of one outcome being greater than the probability of another outcome? We really are moving into the realms of desperate remedies if we have to take these ideas seriously; and still we have not looked at all the quantum interpretations that are on offer. But there are one or two more that I ought to mention before asking whether *any* of our models of reality ought to be taken seriously.

---

[1] Paz and Mahler, *Physical Review Letters* 71 (1993), p. 3235.

COUNSELS OF DESPAIR

If you are looking for unorthodoxy, you need look no further than Roger Penrose. In *The Emperor's New Mind*, he asks (p. 227) the reasonable question 'Is the presence of a conscious being necessary for a "measurement" *actually* to take place?', and gives himself the reasonable answer 'I think that only a small minority of quantum physicists would affirm such a view.' But he then goes on to develop his own variation on the quantum theme by accepting that a particle such as an electron really is spread out in space, instead of being concentrated at a point. People may prefer to think of 'probability' being spread out, rather than the electron itself, says Penrose, but in the experiment with two holes, he claims on p. 252, 'we must accept that the particle "is" indeed in *two* places at once! On this view, the particle *has actually passed through both slits at once*.' His conclusion, though (p. 298) is that 'I believe that the resolution of the puzzles of quantum theory must lie in our finding an improved theory', and he points specifically to the puzzle of non-locality.

Non-locality is the bugbear of most of the other 'reasonable' interpretations of quantum theory that have been put forward. One idea is to give up any notion of describing what happens in an individual quantum process, such as a single photon passing through the experiment with two holes, and to say that quantum mechanics is a purely statistical theory, that only describes what happens in large numbers of measurements of this kind (an 'ensemble'). The ensemble interpretation says that we are allowed to ask what happens when a thousand radioactive atoms (say) are observed after one half-life, and we will get the (correct) answer that half of them will have decayed and half will not. But we are not allowed even to ask the question of what happens to a single radioactive atom observed after one half-life.

This approach may have seemed reasonable decades ago, when quantum physicists could only deal with large numbers of quantum entities – but it looks a little foolish today, when individual photons have been fired through experiments and seen to interfere with themselves. Nevertheless, it is endorsed by (among others) John Taylor, of King's College in London, who says 'any other interpretation is not satisfactory', and specifically that 'I find the many-universes interpretation bizarre. No, I'm sorry, I'm a hard-nosed physicist. Since one has no idea of what goes on in the other universes, they shouldn't be brought in.'[1]

An even more desperate attempt (in my view) to solve the quantum mysteries is an approach pioneered by John von Neumann in the 1930s which says that everyday logic cannot be applied to the quantum world. Everyday logic is called Boolean logic, after George Boole, an Irish mathematician who lived from 1815 to 1864 and was the first person to use a symbolic language and notation to describe purely logical processes. In the mathematical logic that develops from these ideas, terms like 'and' and 'either' are represented by mathematical symbols, and logical arguments can be written out as mathematical equations. The 'quantum-logic' approach to solving the quantum mysteries says that terms like 'and' and 'either' do not have the same meaning in the quantum world as they do in the everyday world, so that offering a photon a choice of either one of two slits to go through takes on a different logical significance. I cannot improve on the comment made by Heinz Pagels, describing the response of a person whose brain is wired up to operate on quantum logic to the puzzles of the quantum world:

> If we tell them about the two-hole experiment they just smile – they have no idea what the problem is. Now we see what the trouble with quantum logic is – it is more restrictive than ordinary Boolean logic. You cannot prove as much with quantum logic, and that is the reason you do not have any sense of weirdness in the physical world. Adopting quantum logic would be like inventing a new logic to maintain the earth was flat if confronted with the evidence that it is round.[2]

A much more intriguing idea is the notion, put forward by John Bell, that there is no difference between the pilot-wave theory and Everett's theory.[3] The heart of Everett's original proposal was that every observer is defined by a quantum 'memory state', and remembers a more or less coherent 'history'. The idea of branching realities creating a multiplicity of parallel universes came later in his argument, and Bell argued that this was an unsuccessful and unnecessary addition. The important point to take on board from Everett, he said, was that we have no access to the past, but only to memories which are themselves

[1] Davies and Brown, *The Ghost in the Atom*, pp. 109, 106.
[2] Pagels, *The Cosmic Code*, p. 180.
[3] Bell, *Speakable and Unspeakable*, Chapter 15.

part of the instantaneous (and therefore non-local!) quantum state of the Universe.

The 'multiplication of universes is extravagant,' says Bell, 'and can simply be dropped without repercussions', while still keeping the notion of a *potential* array of realities described by the wave equation. This is like the pilot-wave theory, in which the wave itself is never localized, or 'reduced', although only one set of variables associated with the wave is 'realized' at any moment in time. Requiring every universe to be real, Bell argues, would be like expecting to find a charged particle at every point in space where there is an elec-tromagnetic field. He preferred to stress the way in which the Everett interpretation describes reality as a distribution of all possible solutions to the quantum wave equation, without any pairing up of different configurations. With no pairing of configurations, then (the theme taken up by Deutsch) there is no 'flow' of time, 'there is no association of the particular present with any particular past', and:

> The structure of the wave function is not fundamentally tree-like. It does not associate a particular branch at the present time with any particular branch in the past any more than with any particular branch in the future. Moreover, it even seems reasonable to regard the coalescence of previously different branches, and the resulting interference phenomena, as *the* characteristic feature of quantum mechanics. In this respect an accurate picture, which does not have any tree-like character, is the 'sum over all possible histories' of Feynman.

But Bell was not writing in support of the many-worlds idea, simply representing it as clearly as he could. He points out that 'Everett's replacement of the past by memories is a radical solipsism – extending to the temporal dimension the replacement of everything outside my head by my impressions ... if such a theory were taken seriously it would hardly be possible to take anything else seriously.' Even Bell, though, could not quite bring himself to dismiss the notion utterly. Later in the same book (p. 194), he says: 'I could almost dismiss it as silly. And yet ... It may have something distinctive to say in connection with the "Einstein Podolsky Rosen" puzzle, and it would be worth-while, I think, to formulate some precise version of it to see if this is really so.' Coming from a man who says (on the same page), 'I never got the hang of complementarity, and remain unhappy about contradictions,' and who was not afraid to dismiss von Neumann's no-hidden-variables argument as 'silly', this comes very close to being an endorsement of the many-worlds interpretation, even though Bell only ever really endorsed the pilot-wave picture, conceptually clean and simple in its own right, but also bringing the essential non-locality of the quantum world into sharp focus, and thereby highlighting the problem that any truly satisfactory theory still has to solve.

I've come back to this different perspective on the many-worlds interpretation for two reasons. First, because it is still my favourite among the traditional interpretations, and if I were forced to offer a 'best buy' from all the ideas outlined so far this would be it. Secondly, Bell's way of explaining what is really going on in the many-worlds version of reality (and to a lesser extent Deutsch's attempts to develop these ideas) brings out clearly the role of time in determining our understanding of the quantum world. There is something very tricky about time, and this trickiness is intimately linked with the nature of quantum reality, and the problem of reconciling the equations of quantum mechanics with those of the everyday world.

This has led to a completely different approach to resolving the quantum puzzles, one which, in effect, starts out from the laws of the everyday world and tries to probe inward towards some sort of quantum truth. But before I take a look at this new approach, it's worth taking a small detour to look at the interplay between quantum mechanics and relativity theory. Any really good description of the way the Universe works (the sought-for 'theory of everything') will, obviously, have to find some way of putting these two great theories together in a coherent way, but that is not what I want to discuss here. Rather, I want to look at the places where the two theories seem incompatible – or at least, where quantum theory seems incompatible with the *special* theory of relativity.

A RELATIVISTIC ASIDE

Once again, the problem was outlined with crystal clarity by Bell. The key concept of the special theory of relativity is that the Universe and the laws of physics should look the same for all observers, regardless of how they are moving (but remember that we are only dealing with constant velocities, not accelerations, in the special theory). This is known as 'Lorentz invariance', although, as we saw in Chapter Two,

Lorentz was not the only person to investigate these phenomena in the years before Einstein made his contribution. Aspect's experiment tells us that we have to abandon local reality, and that either the Universe 'out there' is not real, or there is some form of faster-than-light communication, Einstein's 'spooky action at a distance', going on. Bell suggested that the 'cheapest resolution' of the puzzle is to go back to the kind of relativity theory that existed before Einstein's version, the theory constructed by people like Lorentz on the assumption that there really was an ether.[1]

According to these ideas, there really is a preferred frame of reference, but our measuring instruments are distorted by motion in just the right way to ensure that we can never detect motion 'through' (or 'relative to') the ether. The value of this way of looking at things is that, because there is a preferred frame of reference, it turns out that although in this preferred frame things *can* go faster than light, in other frames of reference where influences seem to go both faster than light and backwards in time this is a kind of optical illusion. If there is a preferred frame of reference, then clocks in that preferred frame will tick away at a preferred rate of time – both Newton's absolute space *and* his absolute time are restored in one fell swoop. It is only in Einstein's version of relativity, with all Lorentz frames equivalent to one another, that going faster than light also means 'really' going backwards in time.

Bell developed these ideas in an article that became Chapter Nine of *Speakable and Unspeakable in Quantum Mechanics*. He showed how using the pre-Einsteinian idea of a preferred frame of reference, combined with the experimental fact that we do not detect motion relative to this frame of reference, leads to the usual form of the Lorentz transformation equations so that (p. 77) 'it is not possible experimentally to determine which, if either, of two uniformly moving systems, is really at rest, and which moving'. Einstein's theory differs from Lorentz's version, Bell points out, in terms of its philosophy, and its style. The philosophical difference is that because it is impossible to say which of the two moving systems (if either of them) is really at rest and which is really moving, then the terms 'really resting' and 'really moving' have no meaning, and only relative motion is important. The difference in style is that Einstein starts out from the hypothesis that the laws of physics look the same to all uniformly moving

[1] Davies and Brown, *The Ghost in the Atom*, p. 48 *et seq.*

observers, and then deduces the Lorentz transformations in a simple and elegant way, instead of starting out from the experimental evidence and following a longer route to the same destination. Just as the Copenhagen Interpretation, say, and the many-worlds interpretation give the same 'answers' to quantum problems, so Lorentz's version of relativity and Einstein's special theory give the same 'answers' in all practical situations. But they suggest different interpretations of what is going on.

Bell's proposal is either revolutionary or reactionary, depending on your point of view. It is certainly not mainstream opinion among physicists today. As he mischievously pointed out, though, there is at least one way out of the dilemma of non-locality which does not involve going back to pre-Einsteinian relativity. 'You know,' he told Paul Davies, 'one of the ways of understanding this business is to say that the world is super-deterministic.'[1] In other words, absolutely everything is predetermined, including the experimenter's choice of what measurements to make in the Aspect experiment. If free will is a complete illusion, this gets us out of the crisis. But if such a theory were to be taken seriously...

Suggesting that the special theory might not be the best way of looking at the world, after all, ought not to come as too much of a shock, since its very name tells us that it is not the last word in relativity theories. It is incomplete, because, unlike the general theory, it does not deal with accelerated motion, or gravity.

Now, I promised not to go into details of what the general theory involves, but since this is only an aside I will only be bending that promise by mentioning a couple of salient features. The way gravity is described by the general theory is in terms of the curvature of spacetime. Instead of some mysterious action at a distance (called gravity) reaching out from the Sun and holding the Earth in its orbit, we are encouraged to think of the Sun as making a 'dent' in spacetime, like the dent a bowling ball would make if placed on a stretched rubber sheet. By following a line of least resistance through curved spacetime, the Earth is made to orbit around the Sun, like a marble rolling around the dip in the rubber sheet caused by the bowling ball.

In principle, the gravitational influence of the Sun (or anything else) extends forever across the Universe, although the curvature of spacetime caused by the Sun gets smaller and smaller as you move

---

[1] Davies and Brown, *The Ghost in the Atom*, p. 47.

further and further away from it. Changes in the gravitational influence can be produced by jiggling masses about in spacetime, making ripples (analogous to the ripples you would get in the stretched rubber sheet if you jiggled the bowling ball up and down) which spread out at the speed of light; these gravitational waves are a prediction of Einstein's general theory, and have been confirmed to exist by recent studies of a system known as the binary pulsar. In this system two dense stars in orbit around one another are losing so much energy in the form of gravitational radiation that their orbital period is changing measurably. The amount by which the orbit is changing exactly matches the predictions of the general theory. This discovery was seen as so important that the researchers who made it (Russell Hulse and Joe Taylor) received the Nobel Prize for their work, in 1993.

But although gravitational radiation moves at the speed of light, there is a sense in which the gravitational influence of an object seems to be non-local. On the usual picture, the gravitational field extends everywhere in space (everywhere in spacetime) all the 'time'. This may be related to another mystery that has worried scientists, off and on, for many decades – the puzzle of inertia. Out in space, where there is no friction, if you give an object a push it will keep moving in the direction you pushed it, until it receives another push. It takes energy to make an object change its direction, or to make it move faster or to slow it down. This is such an important point that the Lorentz-invariant frames of reference of observers moving at constant velocity are often simply referred to as 'inertial frames'. But how does the object 'know' that its motion is (or is not) changing?

In an almost empty universe, containing only one particle, motion would be meaningless; there would be no reference point against which motion could be measured. As soon as there is another particle in the universe, however, there is something against which motion can be measured. If there were only one particle in the universe, it is hard to see how it could have any inertia at all. Would adding just one more particle 'switch on' the full inertia of the first particle? Or would the inertia increase as more and more particles were added? Nobody knows. But in the Universe as we know it the actual behaviour of real objects – their inertial response to being pushed and pulled – seems to indicate that they are 'measuring' their velocity with respect to the average position of all the matter in the Universe.

This is known as 'Mach's Principle', after the Austrian physicist Ernst Mach (1838–1916), and was a major influence on Einstein when he was developing the general theory of relativity. Ironically, in spite of Einstein's efforts, the general theory does not actually explain Mach's Principle or the origin of inertia; doubly ironically, Mach did not like Einstein's theory, even though he helped to inspire it. The puzzle remains. How does an object that is given a push *instantly* take stock of how that push is going to affect its motion relative to all the matter in the Universe, and respond accordingly? We are back in the spooky realms of action at a distance – not in quantum theory, but in the context of Einstein's own masterpiece, the general theory of relativity!

The special theory of relativity, which forbids faster-than-light communication, is known to be an incomplete theory of the Universe, and as Bell has elaborated it is for all practical purposes the same as Lorentzian theory, which does permit faster-than-light signalling. On the other hand, the general theory of relativity, which is a much more satisfactory all-round theory than the special theory, seems, somehow, to have non-locality built in to its structure. And, as I am sure you have noticed, if there is any truth underlying Mach's Principle, then there *is* a preferred frame of reference in the Universe, whether or not there is a physical ether.

We know that the Universe is expanding, and the preferred frame of reference, specified by the average distribution of all the matter in the Universe, is also the one in which that expansion is proceeding perfectly uniformly in all directions. We also know that the Big Bang in which the Universe was born was a hot fireball which filled the Universe with electromagnetic radiation, radiation which has since cooled to become a weak hiss of radio noise with a temperature of just under 3 K (just under − 270°C) still filling the Universe today – the famous cosmic microwave background radiation. An observer is also at rest in the Universe's preferred frame of reference if the observer is not moving relative to the cosmic background radiation. Light itself (in the sense that all electromagnetic radiation can be called 'light') provides us with a preferred frame of reference.

The plot is thickening nicely, and I'll return to some of these themes later. But first, that new way of looking at the old puzzles of quantum mechanics.

AN EXPERIMENT WITH TIME

The nature of time is fundamental to all of the scientific understanding of the world. In quantum physics, the 'unmeasured' state of the Universe is a superposition of all possible states, and the physics has to take account (in principle) of *all* those states. In the modern version of many-worlds theory, developed by Deutsch and others, there is no branching of universes because all the possibilities 'always' exist – there is an infinite number of universes, which 'start out' identical to one another. The process of quantum measurement does not cause a universe to split, but changes the alternative universes in different ways because the outcome of the experiment is different in different universes – in one the cat lives, in the universe next door the cat dies, but in *both* universes there was a live cat before the experiment was carried out (indeed, the two universes were *indistinguishable* until the experiment was carried out). The only sense in which there is an 'arrow of time' in this situation is that some states of the many universes are more complex than others. The complexity – a result of the outcomes of many different quantum measurements – lies in the future, while simplicity lies in the past.

But when many particles are gathered together and allowed to interact, properties related to the complexity of the system (usually described in terms of the branch of physics known as thermodynamics) come into play, and that seems to be where the arrow of time appears. In the classic example, a glass falling from a table shatters when it hits the floor, and the energy it gains in falling is dissipated as heat, warming the floor slightly. We never see the floor giving up energy to the broken bits of glass, pushing them back together and making the whole glass leap back on to the table while the floor cools slightly, even though at the level of atoms and molecules the dynamical equations (both Newtonian and quantum) would 'run' just as well in that direction.

The question of complexity and the arrow of time, and the way in which order seems to emerge out of chaos, has been taken up in particular by Ilya Prigogine. He was born in Russia in 1917 but has lived in Belgium since he was twelve; he won the Nobel Prize for chemistry in 1977, and has since devoted much of his energy to proposing a new interpretation of the way the Universe works. Prigogine has developed mathematical models of non-equilibrium systems, and this work has direct relevance to the origin and evolution of life, and (perhaps) to the puzzle of what happens during a quantum measurement.

The essence of Prigogine's argument is that the experimentally derived thermodynamic laws, based on the behaviour of complex systems, are the true reality, while the apparently time-symmetric behaviour of little spheres bouncing off one another (the naive picture of how atoms behave) is only an approximation to reality. It is the thermodynamic laws, not Newton's laws (or even Schrödinger's equation) which are to be taken at face value. When systems do follow Newton's equations (say) precisely, then they are said to be 'integrable'; the orbit of a single planet around an isolated star is integrable, and as a result the position of the planet can be calculated for any time in the future or any time in the past, provided the present parameters describing the orbit and the position of the planet are known. But add just one more object to the system, creating a 'three-body' problem, and the equations are no longer integrable.

It isn't just that the mathematical equations get harder to solve when more bodies are involved; they become *impossible* to solve precisely, even in principle. It is possible to make reasonably accurate approximate calculations of where each of the three bodies will be at some time in the future, by working in small steps. What you have to do is to pretend that two of the bodies are still, and work out how the other one will move under their combined gravitational influence. But you only let it move a little bit, before 'holding it still' and working out the next tiny movement of one of the other bodies. Then you repeat the process for the third body, and so on. It is a tedious process, even with the aid of fast computers, and it is not perfectly accurate. It actually works quite well for the orbits of planets in the Solar System, but that is because the Sun is so much bigger than any of the planets (indeed, much bigger than all the planets put together), so that its influence dominates the calculation. If each of the planets were the same mass as the Sun, the calculations would be far more difficult to carry out, even approximately. You would get different 'answers' depending upon the order in which you allowed each 'planet' to move a little bit while holding the others still. There is, in fact, no way of predicting precisely how the orbits of three bodies (let alone those of a system as complex as our Solar System) will evolve as time passes – and, similarly, there is no way of calculating exactly how those orbits evolved from the past to get to be the way they are today.

And that is true even when only three interacting objects are involved. Remember that there are 10^{26} 'particles', not just three, in a cat! The beautiful, time-symmetric equations of quantum theory might apply to two or three entities involved in an interaction, but according to Prigogine 'non-integrability' is a fundamental feature of any realistically complex system. And if something is not integrable, then it *cannot* be wound backwards in time to retrace its past, even in principle. The broken glass *cannot* be reconstructed, even if the atoms in the floor do conspire to pump energy into it while they cool down.

In a sense, Prigogine's approach brings back into consideration something like the original Copenhagen Interpretation. The things that matter in physics, he says, are the measurements we make using real, 'classical' apparatus (Geiger counters and the rest), while what goes on inside the apparatus is something we can only ever understand in approximate terms. As Alastair Rae puts it:

> By definition we have no experience of reversible, pure quantum 'events' that are not detected. ... The laws of classical physics were set up on the unquestioned assumption that, although events may be reversible, it is always possible to talk about what has happened. Even Einstein's theory of relativity refers extensively to the sending of signals which are clearly irreversible measurement-type processes. Perhaps it should not be surprising that, when we try to construct a scenario that goes beyond the realm of possible observations into the reversible regime, our models involve apparent contradictions such as wave–particle duality and the spatial delocalization observed in EPR experiments.[1]

These are intriguing and new ideas, which are far from being universally accepted but which are certain to be the subject of debate and development, one way or the other, over the next decade or so. The emphasis is on the possibility that what matters in the Copenhagen Interpretation – the point at which a 'measurement' decides which way the quantum world will jump – is the establishment of an irreversible change (such as the death of a cat, but more prosaically the mark made by a pen on a recording device) in the Universe. The snag, it seems to me, is that there is still no really satisfactory explanation here of the non-locality observed in those EPR experiments that Rae refers to. It's all very well saying that we should expect to be surprised by the quantum world, but there is no inkling of non-locality

[1] Rae, *Quantum Physics*, p. 109.

per se in Prigogine's approach, while non-locality (revealed in the experiment with two holes and the Aspect experiment) is at the very heart of the quantum mysteries. As the Nobel prize-winning physicist Brian Josephson, of Cambridge University, has remarked, the experimental proof that Bell's inequality is violated in the real world is the most important development in physics in recent times.[1] However you dress up the description of the act of measurement, it is still true that the act of measurement on photon A *instantaneously* determines the state of photon B, possibly on the other side of the Universe.

So Prigogine's approach is still not my 'best buy'. But I do agree that the question of reversibility, and the time-symmetric behaviour of some of the fundamental equations, is of key importance in developing a good understanding of quantum reality. And Rae quotes another particularly apposite comment by Prigogine: 'An elementary particle, contrary to its name, is not an object that is "given"; we must construct it.'[2]

The point is that everything we 'know' about the quantum world is based on inferences and observations of things in the everyday world. Physicists deal in models, which are approximations to (they hope) some underlying reality. But they often forget to distinguish between those models and reality itself, while our preconceptions and cultural influences colour the very way in which we begin to think about the way the world works. In order to appreciate what it is that we really do understand about the quantum world (if anything), we ought really to try to understand what we mean by 'understanding' itself. Don't worry; I do not intend to delve into the murky depths of mysticism, philosophy and psychology. But it is still worth looking at how we think about things, in the broadest terms, before we try to assess the various quantum realities on offer to determine just what *is* the best buy, and why.

[1] Quoted in Davies and Brown, *The Ghost in the Atom*, p. 45; full source not cited.
[2] Quoted in Rae, *Quantum Physics*, p. 109; full source not cited.

CHAPTER FIVE

Thinking About Thinking About Things

The physicists' world is made of photons. This is true at two levels. First, everyday things are made of atoms. In order to understand our immediate surroundings and the workings of our own bodies, we do not need to worry unduly about more subtle entities. But an atom is almost entirely empty space, held together by electromagnetic forces – by the exchange of photons. A typical atomic nucleus, carrying positive charge, is about 10^{-15} of a metre across, while the atom itself is a hundred thousand times bigger, 10^{-10} of a metre across. If the nucleus were just 1 cm in diameter, the distance from the nucleus to the outer shell of electrons surrounding it would be a full kilometre. The outer face of the atom, the part that interacts with other atoms, is pure electricity – electrons – held in place by electromagnetic forces (the exchange of photons) in accordance with the workings of QED.

The computer on which I am typing these words seems like a solid object to me; but it is really a web of electromagnetic forces connecting a few tiny, widely-spaced quantum entities – a framework of interacting photons. But what do I mean when I say that I 'feel' the computer to be a solid object, or 'see' it as a continuous entity?

When we feel anything – when I hit the keys on my computer so that my fingers feel a reaction – what we actually feel are the interactions between the electron clouds in the thing and the electron clouds in our fingertips. These are electromagnetic interactions involving photons. When we look at things, obviously we see them using interactions between photons and the atoms (or rather, the electrons in the outer parts of those atoms) in our eyes. When messages about what we feel and see (or hear, or smell, or taste) are passed to our brains, they are passed along networks of nerves using electrical impulses. As we have seen, these nerve impulses cross gaps called synapses, and to do so they stimulate chemical reactions. But chemical reactions themselves are simply processes that involve the electrons in the outer parts of atoms, and which are driven by the quantum processes of electromagnetism. The very workings of our brains themselves depend on the same kind of chemistry – that is, on the exchange of photons.

Even with these limitations, human senses are inadequate to probe the quantum world within the atom. Particles are not seen directly, or tasted, smelled, or touched; rather, their interactions are monitored with the aid of more or less complex machines, and their properties are inferred from readings on a dial, or tracks in a photograph, or numbers counted by a computer. Even when we say that it is now possible to 'see' an individual atom trapped in a magnetic field, what we really mean is that we can see light of the right colour, coming from the right place, to be explained in terms of the presence of the kind of entity we call an atom, whose structure is inferred from many experiments and observations using machines of one kind or another as aids to our senses. The kind of entity that we call an atom is really a theoretical *model* of reality. All the things I have talked about as making up an atom – positively charged nucleus, electron cloud, photons being exchanged – are part of a self-consistent story which both explains past observations and makes it possible to predict what will happen in future experiments. But our understanding of what an atom 'is' has changed several times in the past hundred years or so, and different images (different models) are still useful in different contexts today.

The very name 'atom' comes from the ancient Greek idea of an ultimate, indivisible piece of matter. But by the end of the nineteenth century it had been shown that atoms were not indivisible, and that pieces (electrons) could be knocked off them. Later, a model was developed which described the atom in terms of a nucleus at the centre with the electrons orbiting around it rather like the way planets orbit around the Sun. This model still works very well for explaining how electrons 'jump' from one orbit to another, absorbing or emitting electromagnetic energy as they do so and creating the characteristic lines associated with that kind of atom (that element) in a spectrum. Later still, though, the idea of electrons as waves, or clouds of probability, became fashionable (because these ideas could explain otherwise puzzling features of the behaviour of atoms), and to a quantum physicist the older orbital model was superseded. But this does not necessarily mean that atoms 'really are' surrounded by electron probability clouds, or that all other models are irrelevant.

When physicists are interested in the purely physical behaviour of a gas in the everyday sense – for example, the pressure it exerts on the walls of a container – they are quite happy to treat the gas as little, hard 'billiard balls'. When chemists determine the composition of a substance by burning a small sample and analysing the lines in the spectrum produced, they are quite happy to think in terms of the 'planetary' model of electrons orbiting the nucleus. Yet Nick Herbert, who ought to know better, dismisses this model in his book *Quantum Reality*:

> When my son asks me what the world is made of, I confidently answer that deep down, matter is made of atoms. However, when he asks me what atoms are like, I cannot answer though I have spent half my life exploring this question. How dishonest I feel – as 'expert' in atomic reality – whenever I draw for schoolchildren the popular planetary picture of the atom; it was known to be a lie even in their grandparents' day.[1]

But was it a lie? *Is* it a lie? No! No more so, at least, than any other model of atomic reality. Herbert is being too harsh on himself, on those grandparents, and on physicists in general. The planetary model still works entirely satisfactorily within its limitations, as does the billiard-ball model within *its* limitations. All models of the atom are lies in the sense that they do not represent the single, unique truth about atoms; but all models are true, and useful, in so far as they give us a handle on some aspect of the atomic world.

The point is that not only do we not know what an atom is 'really', we *cannot* ever know what an atom is 'really'. We can only know what an atom is *like*. By probing it in certain ways, we find that, under those circumstances, it is 'like' a billiard ball. Probe it another way, and we find that it is 'like' the Solar System. Ask a third set of questions, and the answer we get is that it is 'like' a positively charged nucleus surrounded by a fuzzy cloud of electrons. These are all images that we carry over from the everyday world to build up a picture of what the atom 'is'. We construct a model, or an image; but then, all too often, we forget what we have done, and we confuse the image with reality. So when one particular model turns out not to apply in all circumstances, even a respectable physicist like Nick Herbert can fall into the trap of calling it 'a lie'.

The way in which physicists construct their models of the quantum

[1] Herbert, *Quantum Reality*, p. 197.

world is based on everyday experience. We can only say that atoms and subatomic particles are 'like' something that we already know. It is no use describing the atom as like a billiard ball to somebody who has never seen a billiard ball, or describing electron orbits as like planetary orbits to somebody who does not know how the Solar System works.

Analogies and modelling can even become completely circular processes, as happens when we try to explain the way atoms interact with one another in, for example, a crystal lattice. In such a crystal, the atoms are held in place by electromagnetic forces in a geometrical array. If one atom were to be displaced from its position, it would be pushed and pulled back into place by electromagnetic interactions involving its neighbours. A useful analogy is to imagine that all the atoms are joined to their immediate neighbours by little springs. If one atom is moved out of position, the electromagnetic forces act like imaginary springs, with the springs on one side being stretched, and so pulling the atom back into place, while the springs on the other side are compressed, and therefore push the atom back into place. We seem to have hit on a really good model of the electromagnetic force acting, under these circumstances, like a spring.

But what is a spring? The most common everyday variety of spring is a piece of metal wire bent into a helical or spiral shape. In the spiral form, it may literally be a component of a clockwork mechanism, the physicists' archetypal model of reality, which makes the analogy all the more appealing. When we push the spring, it pushes back; when we pull it, it pulls back. But why? It does so because it is made of atoms held together by electromagnetic forces! The forces we feel when we push and pull on a spring *are* electromagnetic forces. So when we say that the forces between atoms in a crystal are like little springs, what we are saying is that electromagnetic forces are like electromagnetic forces.

Atoms are such a familiar concept that, as this example shows, it is sometimes hard to see this process of modelling at work where they are concerned. It becomes much clearer when we look at how physicists have constructed their standard model of the subatomic world, using analogies which in many cases are not simply derived from the everyday world, but derived second-hand from our everyday understanding of reality. Within the nucleus (which for the purposes of the simple descriptions of the atom could be regarded as like a positively charged billiard ball), we find particles that are, in some senses, 'like' electrons, and forces that operate 'like' electromagnetism. But electrons and electromagnetism are themselves described as being 'like' things in the everyday world – billiard balls, or waves on a pond, or whatever. Reality is what we make it to be – as long as the models explain the observations, they are good models. But is it really true that electrons and protons were lying in wait to be discovered inside atoms, and that quarks were lying in wait to be discovered inside protons, before human scientists became ingenious enough to 'discover' them? Or is it more likely that essentially incomprehensible aspects of reality at the quantum level are being put into boxes and labelled with names like 'proton' and 'quark' for human convenience?

CONSTRUCTING QUARKS

This question has been addressed by Andrew Pickering, of the University of Edinburgh, in his superb book *Constructing Quarks*. 'The view taken here,' he says in the Preface to that book, 'is that the reality of quarks was the upshot of particle physicists' practice, and not the reverse.' This is what gives him the title of his book, which I have borrowed for this summary of his arguments.

The standard model of reality to which most physicists subscribe sees the everyday world as being made up of essentially four kinds of particle and four kinds of force. The overall picture is complicated slightly because the particles (though not the forces) seem to be duplicated twice, making three 'generations' with closely related properties but different masses. But as far as ordinary atoms are concerned, the four 'first-generation' particles are sufficient to explain everything. The electron itself is one of these 'fundamental' particles, and associated with the electron is a particle called the neutrino. Together, they are dubbed 'leptons'. Protons and neutrons, the particles 'inside' the nucleus, are not, however, thought to be truly fundamental. Instead, they are seen as composed of quarks. Quarks *are* thought to be fundamental, and in the first generation (the counterparts to the two first-generation leptons) they come in two varieties, called 'up' and 'down'. The names have no significance; they are simply labels used by physicists; the two kinds of quark might just as well have been called, say, 'Alice' and 'Bob'.

According to the standard model, a proton is composed of two up quarks and one down quark, held together by one of the four fundamental forces, while a neutron is composed of two down quarks and one up quark, held together in a similar fashion. Because each up quark carries (among other properties) a positive charge equal to two-thirds of the electric charge of an electron, while each down quark carries a negative charge one-third the size of the charge on the electron, the proton ends up with one unit of positive charge ($\frac{2}{3} + \frac{2}{3} - \frac{1}{3} = 1$) while the neutron has no overall charge ($\frac{2}{3} - \frac{1}{3} - \frac{1}{3} = 0$) and is electrically neutral.

As well as the strong force that holds quarks together to make protons and neutrons, and holds protons and neutrons together to make nuclei, there is a weaker nuclear force (known, logically enough, as 'the weak force') which is responsible for radioactivity. The other two fundamental forces are gravity and electromagnetism. Quarks are 'like' electrons, and the strong and weak forces are 'like' electromagnetism, operating by the exchange of particles called bosons which are 'like' photons. It is in many ways a simple and appealing picture, and one which has certainly proved very fruitful in terms of making predictions that have been borne out in experimental tests. Newton would surely have approved. But how did physicists construct this model of the subatomic world?

One of the points that Pickering stresses is that no theory is perfect. Indeed, in principle it is possible to dream up any number of theories that will each explain a particular set of experimental facts. One of the ways in which physicists try to winnow out good theories from bad ones is by choosing the ones that explain the most facts with the fewest assumptions – but, as we saw in Chapter Four, that may still leave you with a choice of explanations. Some theories are simply regarded as less plausible than others, and winnowed out in that way. But any mention of plausibility implies making a judgement – and, again, the example of quantum interpretations shows just how personal and loaded even scientific judgements can be. Most crucially of all, though, throughout the history of science no single theory has ever been able to explain all the facts. Many physicists claim that they are searching for such a theory of everything, or TOE; but if history is any guide their search will be ultimately unsuccessful. There are always areas of disagreement between theory and experiment, and once again there is an element of subjective choice in deciding which of these misfits can be tolerated and which ones spell the downfall of a particular theory.

For, of course, a misfit between experiment and theory may arise because the experiment itself is fallible. The way scientists interpret the outcome of an experiment (especially the kind of experiment used to probe the structure within the proton) depends in large measure upon their theoretical understanding of how the experiment works, and any imperfection of theory may be reflected in an imperfection in the experiment (or at least, our understanding of the experiment) itself. Then again, physicists have to decide just what it is they are measuring. As Pickering points out, in a discipline like particle physics there is the perennial problem of background 'noise'. There are other events going on that can mimic the kind of effects that the experimenters are trying to observe, and which have to be eliminated, if at all possible. This is just like the need to eliminate the background noise (or 'static') picked up by a radio receiver and to tune the receiver in accurately to the signal you want to hear – indeed, physicists refer to the property they are trying to investigate as the 'signal', just as they refer to background interference as 'noise'. It is impossible to eliminate every trace of noise, so again a subjective judgement has to be made about when the experiment is 'good enough' for the required purposes, and the remaining noise can be ignored.

But success breeds success. Once a theory turns out to be (or is perceived to be) a 'good' description of the way things are, it pushes out rival theories, which no longer receive attention. This happened with theories of light. After Newton, the particle theory reigned supreme for a century; after the work of Young and Fresnel, and then Maxwell, the wave theory pushed out the particle theory. Yet *both* theories, we now recognize, are good models. Quark theory is not yet so sophisticated as the wave–particle theory of light. 'By interpreting quarks and so on as real entities,' says Pickering, 'the choice of quark models ... is made to seem unproblematic: if quarks really are the fundamental building blocks of the world, why should anyone want to explore alternative theories?' – even if, as may well be the case, alternative theories can also explain all the experimental results.[1] Many physicists are in grave danger of forgetting that the standard model is just that – a *model*. Protons behave *as if* they contain three quarks; but that does not 'prove' that quarks 'really exist'.

---

[1] Pickering, *Constructing Quarks*, p. 7.

As William Poundstone put it in his book *Labyrinths of Reason*, published in 1988:

> Scientists must be wary of nonprojectable terms. Quarks are hypothetical entities said to reside deep inside protons, neutrons and other subatomic particles. Quarks are counterfactual: Not only has an isolated quark never been observed, but (under most theories) an isolated quark is impossible. Quarks are what a proton *would* split into, *if* it could be split, which it *can't* ... some wonder if [the supposed properties of quarks] may be artificial complications of a simple reality we do not yet understand. Possibly someday someone will hit on how things really are, and we will realize that our current physics is a strained way of describing this reality ... The answer is not in the sky, but in our heads.

But Poundstone is only halfway to appreciating what physics is all about. He does not make the connection with the fact that protons, neutrons and other subatomic particles are also hypothetical entities projected into reality out of our heads by our models. Yes, there may be a simpler way of modelling what goes on at the level of physical phenomena now conventionally explained in terms of the quark model; but that would not be the way things 'really' are, just another model of reality, in the same way that Maxwell's wave equation and Einstein's photons are both good models of the reality represented by the phenomenon of light, and the billiard-ball model and the 'planetary' model of the atom are both good models, depending on which problem you are trying to solve.

The whole of physics, as I have explained, is based upon the process of making analogies and making up models to account for what is going on in realms that we cannot probe with our own senses. The enormous progress that was made in developing the standard model of the particle world in the 1960s and 1970s stemmed from two key analogies, one which took the model of the nucleus as being made of protons and neutrons and carried this over to a model of protons and neutrons as being made of quarks, and one which took the explanation of the electromagnetic force in terms of the exchange of photons and carried this over to a description of the inter-quark forces in terms of the exchange of photon-like particles. The analogy with quantum electrodynamics (QED) is so precise, and so deliberate, that the standard theory of this strong (or 'colour') interaction is given the name quantum chromodynamics, or QCD – 'chromo' because some of the particles involved are given labels that are the names of colours, an arbitrary convention like the naming of the 'up' and 'down' quarks, and not an indication that the particles are 'really' coloured in the everyday sense of the term.

But the quark theory did not leap into consideration, fully fledged, and sweep all opposition aside in one fell swoop. It crept up on physicists, almost against their better judgement. Two theorists came up with the idea independently at about the same time, in the early 1960s, but neither of them went out on the campaign trail for the new theory. One of the proposers of quark theory was Murray Gell-Mann, an American physicist (he was born in New York City in 1929) who actually came up with the name, borrowing it from a line in James Joyce's *Finnegans Wake*. He was working at the California Institute of Technology, and had already established a formidable reputation as one of the greatest theorists around. He had been involved in successful attempts to group the particles known to physicists according to their properties, and to make predictions about the properties of particles that were yet to be discovered, in much the same way that Dmitri Mendeleev had grouped the chemical elements together in the periodic table and made predictions about the properties of yet-to-be-discovered elements in the nineteenth century – another example of the power of analogy, and the traditionalist nature of science.

This search for patterns also led to the realization that many of the properties of protons and neutrons could be explained in terms of fundamental triplets (what we now call quarks) arranged in different ways. Gell-Mann published the idea almost ashamedly, in 1964, in a paper only two pages long, in the journal *Physics Letters*. One reason for his hesitancy – and for the reluctance of most physicists to take the quark idea seriously for several years – was the suggestion that the triplet particles ought to have electric charges a fraction of the charge on the electron, which was at that time firmly established as the 'smallest possible' unit of charge. Today, nobody worries about quarks having charge two-thirds or one-third as big as the charge on the electron, but in 1964 'everybody knew' that this was impossible. So Gell-Mann concluded his paper by almost disowning his own proposal. He suggested, in effect, that the triplets that could so neatly explain properties of protons and neutrons were really just mathematical devices, a way to get a handle on some of the properties of protons and neutrons, and summed up:[1]

---

[1] *Physics Letters* 8 (1964), p. 214.

It is fun to speculate about the way quarks would behave if they were physical particles of finite mass (instead of purely mathematical entities as they would be in the limit of infinite mass) ... A search for stable quarks of charge $-\frac{1}{3}$ or $+\frac{2}{3}$ and/or stable diquarks of charge $-\frac{2}{3}$ or $+\frac{1}{3}$ or $+\frac{4}{3}$ at the highest energy accelerators would help to reassure us of the non-existence of real quarks.

Even the theorist who 'invented' quarks wanted to be reassured that they were a figment of his imagination, and did not really exist! This is not quite so bizarre as it sounds at first. Gell-Mann's approach to the 'discovery' of quarks was highly mathematical and rather esoteric; he found that certain features of the equations could be explained by treating protons and neutrons *as if* they were composed of triplets, but he started from the mathematical end, not from considering these triplets to be physically real particles.

The other theorist who invented quarks was slightly less ambiguous about his creation, but found that promoting the idea was not the best way to further his career. George Zweig had been born in Moscow in 1937, but moved to the United States, where he completed a BSc at the University of Michigan in 1959, and then began research at CalTech. He started out as an experimenter, but after three unsuccessful years switched to theory, working for his PhD under the supervision of Richard Feynman. Like Gell-Mann, he realized that the properties of particles such as protons and neutrons could be explained by treating them as triplet composites of other particles, which he called 'aces'. But, perhaps because he was relatively young and new to the particle-physics game (and therefore less hidebound by tradition), he was more inclined to throw caution to the winds and accept these entities as physically real particles.

In 1963, Zweig moved to CERN, the European centre for particle research in Geneva, where he completed his thesis and wrote the discovery up for publication. His 'discovery' paper was also published in 1964, and in that paper Zweig concluded that 'in view of the extremely crude manner in which we have approached the problem, the results we have obtained seem somewhat miraculous'.[1]

Most of the physics community seemed to agree, and rather than receiving recognition for his insight Zweig was on the edge of being

[1] CERN Preprint, number 8182/TH401.

tarred with the brush of crankiness. In 1980 he told an international conference that:

> The reaction of the theoretical physics community to the ace model was generally not benign. Getting the CERN report published in the form that I wanted was so difficult that I finally gave up trying. When the physics department of a leading university was considering an appointment for me, their senior theorist, one of the most respected spokesmen for all of theoretical physics, blocked the appointment at a faculty meeting by passionately arguing that the ace model was the work of a 'charlatan'.[1]

And this was in spite of the fact that in his original draft of what became the CERN 'preprint', running to 24 pages, Zweig had, unlike Gell-Mann, properly set out the consequences of the triplet idea, in great detail.

The injustice did not end there. Gell-Mann received the Nobel Prize for physics in 1969, for his overall contributions and discoveries concerning the classification of elementary particles and their interactions. The award was undoubtedly justified, but even by 1969 quark theory had not quite become established and the citation for the award did not specifically mention that work. By the time quark theory became established as part of the standard model, it would have been inappropriate to give Gell-Mann a second bite of the cherry, and presumably the Nobel Committee felt that they could not give Zweig recognition without including Gell-Mann as well. It remains a glaring anomaly that the man who first suggested that quarks might be real, and who first spelled out the implications in detail, pointing the way to what has become the standard model of particle physics, has not received a Nobel Prize. But still, nobody ever claimed that Nobel Prizes are awarded entirely fairly and solely on rational grounds.

The quark theory only began to be taken more seriously when experiments involving collisions between particles (electrons being bounced off protons, and protons being bounced off one another) began to show up structure inside the proton. But this is not quite so simple as it seems at first sight, because protons are intrinsically more complex creatures than electrons, whether or not they are made of quarks (the following discussion of internal structure applies to neutrons, as well, but the experiments are actually carried out with protons because their electric charge provides a handle by which they can be

[1] In *Baryon '80*, ed. by N. Isgur (Toronto: University of Toronto Press, 1981), p. 439.

tugged by magnetic fields and accelerated to high energies).

Remember the way in which an electron is regarded, in QED, as a point surrounded by a cloud of 'virtual' photons, electron–positron pairs, and the rest. The magnetic moment of the electron is calculated with exquisite precision by taking account of ever more complicated ('higher order') possibilities of this kind, but each step upward in complexity provides a smaller correction to the calculation. Because it carries positive charge, the proton will also be involved in electromagnetic interactions of this kind, and has a magnetic moment which can be calculated in the same way as the electron's magnetic moment. But unlike the electron the proton also 'feels' the strong force. Even before theorists realized that the strong force involves interactions between quarks, they knew that it operated between protons and neutrons to hold the nucleus together, and they were able to investigate some of its properties. By analogy with QED, they found that a proton should be surrounded by a cloud of other particles, including proton–antiproton pairs, neutron–antineutron pairs, and force-carriers called mesons (the strong field's equivalent of photons). But there is a crucial difference. In the strong field, these extra influences do not get smaller as things get more complicated – at higher orders of the calculation. Instead of the higher order influences being a small correction, they are as important as the 'real' proton itself. The result is that according to quantum field theory, the proton has to be considered as a complex ball of interacting virtual particles, extending over the range of the strong force – which, fortunately, is only about 10^{-13} cm.

Experiments to probe the structure of the proton depended on having a good theory of the electron – QED itself. It was only because theorists were confident that they understood electrons, and that electrons really could be treated as point-like objects, that they could interpret the way electrons 'scattered' off protons to reveal the structure of the protons themselves. When high-energy (that is, fast-moving) electrons are bounced off each other in accelerator experiments, they tend to be scattered at very large angles, ricocheting off one another as if they were hard objects, like billiard balls. When electrons are bounced off protons, however, they are usually deflected by only small angles, as if they are scattering off a soft object which can only give them a gentle nudge. The two kinds of interaction are known as 'hard' and 'soft' scattering experiments. These experiments showed that protons did indeed have a diameter of about 10^{-13} cm, and gave a great boost to the field theorists. But the 'answers' Nature gave to the questions posed by the experimenters still depended on their choice of which experiments to carry out, and what to measure. As the philosopher Martin Heidegger has put it:

> Modern physics is not experimental physics because it uses experimental devices in its questioning of nature. Rather the reverse is true. Because physics, already as pure theory, requests nature to manifest itself in terms of predictable forces, it sets up the experiments precisely for the sole purpose of asking whether and how nature follows the scheme prescribed by science.[1]

In a curious echo of the way in which two people had discovered (or invented) quarks by two very different routes earlier in the decade, late in the 1960s two field theorists developed equally different approaches to an explanation of the detailed results of scattering experiments. One of them, James Bjorken, at Stanford University, followed the equivalent of Gell-Mann's approach, from the mathematical end. His explanations of these phenomena worked, mathematically speaking, but according to Pickering 'they were esoteric to the point of incomprehensibility'.[2] The other approach, however, came from Richard Feynman, who could be relied upon to be both insightful *and* comprehensible.

The great thing about Feynman's approach was that it made sense to physicists brought up in the tradition of taking things (like atoms) apart to find out what they are made of. He developed his ideas in the mid-1960s, and published them in 1969. Without prejudging the issue of whether or not quarks existed, he developed a *general* explanation of what happens when a high-energy electron probes inside a proton, or when two high-energy protons collide head on.

Feynman's starting point was the field-theory idea that a proton must be a swarm of particles. The strict analogy with QED said that it should be a ball of protons, neutrons and their antiparticles, plus mesons; quark theory said that it should be composed of three basic quarks, but that *they* would each be associated with a cloud of their own virtual particles. Deliberately agnostic, Feynman gave these inner components of the proton the name 'parton', to cover both possibilities.

---

[1] Heidegger, *The Question Concerning Technology* (New York: Harper & Row, 1977), p. 21.
[2] Pickering, *Constructing Quarks*, p. 132. I am glad to have Pickering's view on this, since they are certainly incomprehensible to me, and I only have the word of my mathematical friends that they really do work.

But he realized that very little of this complexity mattered in a single collision. When an electron is fired into a proton, it may exchange a photon with a single parton, which recoils as a result while the electron is deflected, but that is the limit of its influence on the proton (and of the proton's influence on the electron). Even if two protons smash into each other head on, what actually happens is that individual partons from the two protons interact with one another in a series of point-like hard scattering events. Bjorken's calculations showed (to the cognoscenti!) that a particular mathematical framework could explain the way protons scattered, and then said, almost as an afterthought, that one way of arriving at this mathematical framework would be if protons contained point-like particles; Feynman said that if protons contained point-like particles then this would lead to a mathematical description that explained the observations of scattering.

Pickering argues that the reason why Feynman's approach swept the board, and led to further experiments which established the 'reality' of quarks to the satisfaction of most theorists, is that it follows a long-established and well-understood tradition. Theorists had a classical analogy ready and waiting, in the form of the experiments early in the twentieth century that had probed the structure of the atom. The pioneering particle physicist Ernest Rutherford bombarded atoms with so-called alpha particles (now known to be helium nuclei), and found that some of the alpha particles were scattered at large angles, showing that there was something hard and billiard-ball-like in the centre of an atom (its nucleus). Experiments in the 1960s showed electrons sometimes being scattered at surprisingly large angles from within otherwise 'soft' protons, and Feynman's model explained this in terms of hard, billiard-ball-like entities within the proton.

It took years for the standard model to become established, but once the physicists had been set thinking along these lines there was an air of inevitability about the whole process. With two great analogies to draw on – the nuclear model of the atom, and the QED theory of light – the quark model of protons and neutrons and the QCD theory of the strong interaction became irresistible. 'Analogy was not one option amongst many,' says Pickering; 'it was the basis of all that transpired. Without analogy, there would have been no new physics.'[1]

The same is true of quantum mechanics itself. Indeed, it is hard to see quantum physics as anything but analogy – the wave–particle

---

[1] Pickering, *Constructing Quarks*, p. 407.

duality being the classic example, where we struggle to 'explain' something we do not understand by using *two*, mutually exclusive, analogies which we apply to the same quantum entity.

But Pickering also raises another intriguing, perhaps disturbing, question. Was the path to the standard model of particle physics inevitable? Is this the real (or only) truth about the way the world works? None of the theories which led up to the standard model was ever perfect, he points out, and particle physicists continually had to choose which theories to give up and abandon and which ones to develop to try to make a better fit with experiment. The theories that they chose to develop also influenced the choices of which experiments to carry out, and this interacting chain of decisions led to the new physics. The new physics was a product of the culture in which it was created.

The philosopher of science Thomas Kuhn has carried this kind of argument to its logical conclusion, arguing that if scientific knowledge really is a product of culture then scientific communities that exist in different worlds (literally on different planets, perhaps; or at different times on the same planet) would regard different natural phenomena as important, and would explain those phenomena in different theoretical ways (using different analogies). The theories from the different scientific communities – the different worlds – could not be tested against one another, and would be, in philosopher's jargon, 'incommensurable'.

This runs counter to the way most physicists think about their work. They imagine that if we ever make contact with a scientific civilization from another planet then, assuming language difficulties can be overcome, we will find that the alien civilization shares our views about the nature of atoms, the existence of protons and neutrons, and the way the electromagnetic force works. Indeed, more than one science-fiction story has suggested that science *is* the (literally) universal language, and that the way to set up communication with an alien civilization will be by describing, for example, the chemical properties of the elements, or the nature of quarks, to establish a common ground. If the aliens turn out to have completely different ideas about what atoms are, or to have no concept of atoms at all, that would make such attempts at finding common ground doomed from the start.

The idea of science as a universal language is usually most forcefully expressed in terms of mathematics. Many scientists have commented on the seemingly magical way in which mathematics 'works' as a tool for describing the Universe; Albert Einstein once said that 'the most incomprehensible thing about the Universe is that it is comprehensible'. I have sometimes puzzled over the fact that it is possible for an ordinary human being to learn enough about the Universe, in less than a human lifetime, to 'comprehend' it in this way. But now I think that this is not such a mystery after all. Pickering persuades me that I was looking at the puzzle through the wrong end of the telescope. He quotes John Polkinghorne, a British quantum theorist who is also a Minister in the Church of England, as saying that 'it is a non-trivial fact about the world that we can understand it and that mathematics provides the perfect language for physical science; that, in a word, science is possible at all'.[1]

But such assertions, says Pickering, are mistaken:

> It is *unproblematic* that scientists produce accounts of the world that they find comprehensible: given their cultural resources, only singular incompetence could have prevented members of the [physics] community producing an understandable version of reality at any point in their history. And, given their extensive training in sophisticated mathematical techniques, the preponderance of mathematics in particle physicists' accounts of reality is no more hard to explain than the fondness of ethnic groups for their native language.

In other words, the 'mystery' that mathematics is a good language for describing the Universe is about as significant as the discovery that English is a good language for writing plays in. If world views really are cultural products, as Pickering and Kuhn argue, then it should be no surprise that there are different interpretations of quantum reality. But before I develop this theme further, perhaps a couple of examples from another area of science will help to persuade you that it really is no surprise that we can describe the Universe using mathematics, and that how we interpret that mathematical description of reality is in large measure (perhaps entirely) a matter of choice.

---

[1] This and the next quote from Pickering, *Constructing Quarks*, p. 413.

PUTTING EINSTEIN IN PERSPECTIVE

One example of the power of mathematics in describing the world, which I have often used myself, is the way in which what had seemed to nineteenth-century mathematicians to be abstract geometrical ideas without any relevance to the real Universe turned out to provide the cornerstone of Albert Einstein's general theory of relativity. One of the entertaining twists in the tale is that Einstein himself did not realize this at first, and had to have the power of the nineteenth-century approach drawn to his attention rather forcefully before he saw the light and was able to use the mathematics to develop his model of the world.

The key feature of Einstein's general theory of relativity is the idea of bent spacetime. But Einstein was neither the originator of the idea of spacetime geometry, nor the first to conceive of space being bent. The easy way to understand Einstein's two theories of relativity is in terms of geometry. Space and time, as we saw in Chapter Two, are part of one four-dimensional entity, spacetime. The special theory of relativity, which deals with uniform motions at constant velocities, can be explained in terms of the geometry of a flat, four-dimensional surface. The equations of the special theory that, for example, describe such curious phenomena as time dilation and the way moving objects shrink are in essence the familiar equation of Pythagoras' theorem, extended to four dimensions, and with the minor subtlety that the time dimension is measured in a negative direction.

Once you have grasped this, it is easy to understand Einstein's general theory of relativity, which is a theory of gravity and accelerations. What we are used to thinking of as forces caused by the presence of lumps of matter in the Universe (like the Sun) are due to distortions in the fabric of spacetime. The Sun, for example, makes a dent in the geometry of spacetime, and the orbit of the Earth around the Sun is a result of trying to follow the shortest possible path (a geodesic) through curved spacetime.

Of course, you need a few equations if you want to work out details of the orbit. But that can be left to the mathematicians. The physics is disarmingly simple and straightforward, and this simplicity is often represented as an example of Einstein's 'unique genius'.

Only none of this straightforward simplicity came from Einstein.

Take the special theory first. When Einstein presented this to the world in 1905, it was a mathematical theory, based on equations. It didn't make a huge impact at the time, and it was several years before the science community at large really began to sit up and take notice. They did so, in fact, only after Hermann Minkowski gave a lecture in Cologne in 1908. It was this lecture, published in 1909 shortly after Minkowski died, that first presented the ideas of the special theory in terms of spacetime geometry. His opening words indicate the power of the new insight:

> The views of space and time which I wish to lay before you have sprung from the soil of experimental physics, and therein lies their strength. They are radical. Henceforth space by itself, and time by itself, are doomed to fade into mere shadows, and only a kind of union of the two will preserve an independent reality.[1]

Minkowski's enormous simplification of the special theory had a huge impact. It is no coincidence that Einstein received his first honorary doctorate, from the University of Geneva, in July 1909, nor that he was first proposed for the Nobel Prize in Physics a year later.

There is a delicious irony in all this. Minkowski had, in fact, been one of Einstein's teachers at the Zurich polytechnic at the end of the nineteenth century. Just a few years before coming up with the special theory, Einstein had been described by Minkowski as a 'lazy dog', who 'never bothered about mathematics at all'. The lazy dog himself was not, at first, impressed by the geometrization of relativity, and took some time to appreciate its significance. Never having bothered much with maths at the polytechnic, he was remarkably ignorant about one of the key mathematical developments of the nineteenth century, and he only began to move towards the notion of *curved* spacetime when prodded that way by his friend and colleague Marcel Grossman.

This wasn't the first time Einstein had enlisted Grossman's help. Grossman had been an exact contemporary of Einstein at the polytechnic, but a much more assiduous student who not only attended the lectures (unlike Einstein) but kept detailed notes. It was those notes that Einstein used in a desperate bout of last-minute cramming which enabled him to scrape through his final examinations at the polytechnic in 1900.

What Grossman knew, but Einstein didn't until Grossman told him,

---

[1] Quoted in Abraham Pais, *Subtle is the Lord* (Oxford: Oxford University Press, 1982), p. 152. Other quotes in this section from the same source.

in 1912, was that there is more to geometry (even multidimensional geometry) than good old Euclidean 'flat' geometry.

Euclidean geometry is the kind we encounter at school, where the angles of a triangle add up to exactly 180°, parallel lines never meet, and so on. The first person to go beyond Euclid and to appreciate the significance of what he was doing was the German Karl Gauss, who was born in 1777 and had completed all of his great mathematical discoveries by 1799. But because he didn't bother to publish many of his ideas, non-Euclidean geometry was independently discovered by the Russian Nikolai Ivanovich Lobachevsky, who was the first to publish a description of such geometry in 1829, and by a Hungarian, János Bolyai.

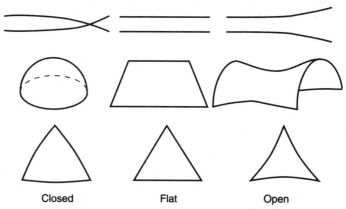

Closed   Flat   Open

Figure 20

Space may conform to one of three basic geometries. We can represent these in two dimensions, even though space is three-dimensional.

If space is positively curved, the Universe is closed. In positively curved space (left), lines that start out parallel (in the usual meaning of the term) may cross, and the angles of a triangle add up to more than 180 degrees.

If space is negatively curved, the Universe is open. In negatively curved space (right) parallel lines can diverge from one another, and the angles of a triangle add up to less than 180 degrees.

If space is flat, parallel lines and triangles obey the rules of geometry that we learned in school ('Euclidean geometry'). Flat space (centre) is a special case on the dividing line between positive and negative curvature. Our Universe is indistinguishably close to being flat.

They all hit on essentially the same kind of 'new' geometry, which applies on what is known as a 'hyperbolic' surface, which is shaped like a saddle, or a mountain pass. On such a curved surface, the angles of a triangle always add up to *less* than 180°, and it is possible to draw a straight line and mark a point, not on that line, through which you can draw many more lines, none of which crosses the first line and all of which are, therefore, parallel to it.

But it was Bernhard Riemann, a pupil of Gauss, who put the notion of non-Euclidean geometry on a comprehensive basis in the 1850s, and who realized the possibility of yet another variation on the theme, the geometry that applies on the closed surface of a sphere (including the surface of the Earth). In spherical geometry, the angles of a triangle always add up to *more* than 180°, and although all 'lines of longitude' cross the equator at right angles, and must therefore all be parallel to one another, they all cross each other at the poles.

Figure 21
A sphere, like the surface of the Earth, is the archetypal closed surface. On a spherical surface, the angles of a triangle can add up to 270 degrees – three right angles.

Riemann gave a lecture on 10 June 1854, under the title 'On the hypotheses which lie at the foundations of geometry'. In that lecture – which was not published until 1867, the year after Riemann died – he covered an enormous variety of topics, including a workable definition of what is meant by the curvature of space and how it could be measured, the first description of spherical geometry (and even the speculation that the space in which we live might be gently curved, so that the entire Universe is closed up, like the surface of a sphere, but in three dimensions, not two), and, most important of all, the extension of geometry into many dimensions with the aid of algebra.

Riemann died, of tuberculosis, at the age of 39, in 1866. But Einstein was not even the second person to think about the possibility of space in our Universe being curved. Chronologically, the gap between Riemann's work and the birth of Einstein is nicely filled by the life and work of the English mathematician William Clifford, who lived from 1845 to 1879, and who, like Riemann, died of tuberculosis. Clifford translated Riemann's work into English, and played a major part in introducing the idea of curved space and the details of non-Euclidean geometry to the English-speaking world. He knew about the possibility that the three-dimensional Universe we live in might be closed and finite, in the same way that the two-dimensional surface of a sphere is closed and finite, but in a geometry involving at least four dimensions. This would mean, for example, that just as a traveller on Earth who sets off in any direction and keeps going in a straight line will eventually get back to their starting point, so a traveller in a closed universe could set off in any direction through space, keep moving straight ahead, and eventually end up back at their starting point.

But Clifford realized that there might be more to space curvature than this gradual bending encompassing the whole Universe. In 1870, he presented a paper to the Cambridge Philosophical Society (at the time, he was a Fellow of Newton's old college, Trinity) in which he described the possibility of 'variation in the curvature of space' from place to place, and suggested that 'small portions of space *are* in fact of nature analogous to little hills on the surface [of the Earth] which is on the average flat; namely, that the ordinary laws of geometry are not valid in them'. In other words, still seven years before Einstein was born, Clifford was contemplating *local* distortions in the structure of space – although he had not got around to suggesting how such distortions might arise, nor what the observable consequences of their existence might be, and the general theory of relativity actually portrays the Sun and stars as making dents, rather than hills, in spacetime, not just in space.

Clifford was just one of many researchers who studied non-Euclidean geometry in the second half of the nineteenth century – albeit one of the best, with some of the clearest insights into what this might mean for the real Universe. His insights were particularly profound, and it is tempting to speculate how far he might have gone in pre-empting Einstein, if he had not died eleven days before Einstein was born.

When Einstein developed the special theory, he did so in blithe ignorance of all this nineteenth-century mathematical work on the geometry of multidimensional and curved spaces. The great achievement of the special theory was that it reconciled the behaviour of light, described by Maxwell's equations of electromagnetism (and in particular the fact that the speed of light is an absolute constant) with mechanics – albeit at the cost of discarding Newtonian mechanics and replacing them with something better.

Because the conflict between Newtonian mechanics and Maxwell's equations was very apparent at the beginning of the twentieth century, it is often said that the special theory is very much a child of its time, and that if Einstein had not come up with it in 1905 then someone else would have, within a year or two.

On the other hand, Einstein's great leap from the special theory to the general theory – a new, non-Newtonian theory of gravity – is generally regarded as a stroke of unique genius, decades ahead of its time, that sprang from Einstein alone, with no precursor in the problems faced by physicists of the day.

That may be true; but what this conventional story fails to acknowledge is that Einstein's path from the special to the general theory (over more than ten years) was, in fact, more tortuous and complicated than it could, and should, have been. The general theory actually follows as naturally from the *mathematics* of the late nineteenth century as the special theory does from the *physics* of the late nineteenth century.

If Einstein had not been such a lazy dog, and had paid more attention to his maths lectures at the polytechnic, he could very well have come up with the general theory soon after he developed the special theory, in 1905. And if Einstein had never been born, then it seems entirely likely that someone else, perhaps Grossman himself, would have been capable of jumping off from the work of Riemann and Clifford to come up with a geometrical theory of gravity during the second decade of the twentieth century.

If only Einstein had understood nineteenth-century geometry, he would have got his two theories of relativity sorted out a lot more quickly. It would have been obvious how they followed on from earlier work; and, perhaps, with less evidence of Einstein's 'unique insight' and a clearer view of how his ideas fitted in to mainstream mathematics, he might even have got the Nobel Prize for his general theory.

That, at least, is one way of presenting the story, emphasizing the power of mathematics. Indeed, this version of the story closely follows an article I wrote for *New Scientist* early in 1993. But as a result of that article I received a communication from Bruno Augenstein, a researcher with RAND, in Santa Monica, California. What Augenstein had to tell me put the story in a rather different light, and helped to convince me that Pickering was right about the way science works.

'For some time,' Augenstein told me, 'I was of the Wigner/Dyson school ("the unreasonable effectiveness of mathematics in the physical sciences ..."), but am now persuaded that the notion you express in your article could be given the status of a strong operational axiom. That is: literally every version of mathematical concepts has a physical model somewhere, and the clever physicist should be advised to deliberately and routinely seek out, as part of his activity, physical models of already discovered mathematical structures.'

In other words, as Pickering has suggested, physicists are capable of producing understandable versions of reality given *any* self-consistent raw material.

I have to admit that it was not just the clarity and insightfulness of my article that persuaded Augenstein that this might be the case. He had already found that a rather obscure branch of mathematics (what he calls 'a somewhat surreal corner of set theory') concerning a subject known as the Banach–Tarski theorems (BTT) provides another example where work in pure mathematics, which was originally thought not to have any relevance to the real world, can now be seen as having anticipated a realization subsequently found in physics. In this case, that realization is the original quark theory of Gell-Mann and Zweig.

I won't go into details; my understanding of 'surreal corners of set theory' is not up to the task, and I have to take Augenstein's word for it.[1] But the essential feature is that the work of Banach and Tarski (which was published in 1924) deals with the way objects can be decomposed into constituent parts and reassembled to make something different.[2] As Augenstein puts it, 'you can cut solid body A, of any finite size and arbitrary shape, into *m* pieces which, without any alteration, can be reassembled into solid body B, also of any finite size and arbitrary shape'.

Surreal indeed – but so general as to be of little practical value. So he has taken up a specific version of this behaviour dealing with solid spheres. In particular, a solid sphere with unit radius can be cut into five pieces in such a way that two of the pieces can be reassembled into one solid sphere with unit radius, while the other three pieces are reassembled into a second solid sphere with unit radius. These are the minimum numbers of pieces required to do the trick, but it can be repeated indefinitely – and perhaps you can guess what is coming next.

---

[1] Reassured by the fact that his ideas have been accepted for publication in *Speculations in Science and Technology*, which in spite of its name is a highly respectable scientific journal.

[2] S. Banach and A. Tarski, *Fundamenta Mathematica* 6 (1924), p. 244.

In the paper he published in *Speculations in Science and Technology*, Augenstein shows that the rules governing the behaviour of these mathematical sets and sub-sets are formally exactly the same as the rules which describe the behaviour of quarks and 'gluons' in the standard model of particle physics, quantum chromodynamics (QCD), which was developed half a century after the original BTT paper appeared; but the physicists who developed the standard model knew nothing of that surreal corner of set theory. Remember that neutrons and protons, in this model, are made up of triplets of quarks, and the gluons that bind protons and neutrons together (equivalent to photons in QED) are made of pairs of quarks.

The magical way in which a proton entering a metal target can produce a swarm of new copies of protons emerging from that target, each identical to the original proton, is precisely described by the BTT process of cutting spheres into pieces and reassembling them to make pairs of spheres. BTT has been described as 'the most surprising result of theoretical mathematics', a view which Augenstein endorses, and which you may well be inclined to agree with.

Intriguingly, Augenstein's analogy also makes predictions. Just as protons were once thought to be featureless billiard balls, and were then probed by high-energy electrons to 'reveal' the three quarks inside (and just as the atom was probed by Rutherford to reveal the nucleus inside) the latest plans of the experimental physicists suggest that it may be possible to go to higher energies still and probe 'within' quarks – if, that is, there is anything within quarks. Interestingly, the five mathematical 'pieces' in Augenstein's version of the Banach–Tarski theorems are a mixture; four of the pieces would imply a very detailed structure within quarks, while the fifth is the mathematical description of a single point.

Augenstein is not the only person to have been intrigued by the implications of the Banach–Tarski theorems for particle physics. In 1982, Roger Jones wrote in his book *Physics as Metaphor*:

Why does the muon exist when it can do no more than what an electron already does ...? The muon is roughly 200 times more massive than the electron ... they differ in [only] one important quantitative measure – mass.

Other particles differ by several significant measures, but the electron and the muon are like two line segments, made in the same way out of the elemental points but having different length. The electron and muon are balls of different size, but with the same number of points. ... Size itself, measure, and number are mere appearance and metaphor, and they should not be mistaken for some ultimate invariant – they should not be idolized.

But in the case of the three-dimensional measure, volume, there is yet another consideration that must give us pause. This is the astounding and paradoxical theorem of Banach and Tarski which says that a sphere of any given size may be taken apart and reassembled into a sphere of a different size ... an electron may be transformed into a muon in a finite number of steps.

To the extent that matter today is represented by some kind of an abstract distribution in a mathematical space ... what we are really talking about is a more organic, unified, chaotic sense of space. It is not a space lacking or deficient in anything, but rather one that is different from ours – another metaphor.

Will physicists take up these ideas and develop a new 'standard model' that goes beyond the quark–QCD description of reality? Or will they languish as a curious byway of science, a bizarre mathematical curiosity regarded as having no physical significance? That remains to be seen. Augenstein likens the physicists' descriptions of reality to fairy tales, and stresses that a significant change in the attitudes and habits of physicists would be required before the idea that models can be constructed almost at will from whatever ingredients are available could be taken seriously. Such a change in attitudes may be a long time coming, if it ever happens; but it strongly echoes Pickering's conclusions about the way physicists come up with their models, and some other physicist–philosophers are already developing these ideas further, looking at where models come from, and how physicists take hold of the world. Indeed, some physicists may already be moving in the direction pointed to by Pickering and Augenstein, without quite realizing the significance of what they are doing.

DESCRIBING THE UNDESCRIBABLE

Having put a toe into the murky waters of the more surreal pools of set theory, I'd like to give just one brief example from cosmology before getting back to the philosophers' view of what physics is all about.

Like the particle physicists with their quarks and QCD, 'explaining' how things work on the small scale, cosmologists have a standard model of how the Universe works on the large scale, involving matter, gravity and the general theory of relativity. One of the big problems – perhaps *the* big problem – with the cosmologists' standard model, the Big Bang theory, is the presence of a singularity at the birth of the Universe. Astronomers know that the Universe is expanding because their telescopes show that galaxies are moving apart from one another. Einstein's general theory of relativity predicted this expansion, because the theory says that the space between the galaxies must be stretching as time passes. Both theory and observation suggest that if you imagine winding this expansion backwards in time to find out what the Universe was like long ago, you must reach a moment in time when all of the matter and all of the spacetime in the Universe was concentrated in a single point, the singularity.

A singularity is a place where the laws of physics as we know them break down. Taking the equations literally, it is a point of zero volume and infinite density, which seems absurd. Yet in the 1960s Stephen Hawking and Roger Penrose showed that if the general theory of relativity is an accurate description of the way the Universe works (which it certainly seems to be in the light of all the evidence, including that of the binary pulsar), then there is no escape from the requirement of a singularity at the beginning of time. The kind of expansion that we observe around us today, coupled with Einstein's equations, proves that there must have been a singularity in the beginning.

But is this disturbing conclusion simply the result of making the wrong kind of analogy? In the 1980s, Hawking returned to the puzzle of the origin of the Universe, and, with others, attempted to find a way to describe the Universe in a model which incorporates the ideas of quantum mechanics, as well as those of the general theory of relativity. This is the work which leads many cosmologists to feel that some variation on the 'many-worlds' or 'many-histories' idea is required, because there is no way an observer can be 'outside' the Universe to collapse its wave function from a superposition of states into a unique history. But there is another intriguing feature of Hawking's approach, a new analogy which gives a different perspective on the Big Bang.

I said before that there is an important difference in the way time and space are treated by the equations of relativity (both the special and the general theories). Time, I mentioned, appears in the equations with a minus sign in front of it. But this is not quite the whole story, because those equations also deal, like Pythagoras' famous theorem about right-angled triangles, in *squares*. So the parameters that represent spatial displacements in Einstein's equations are squares: x^2, y^2 and z^2. The parameter that represents temporal displacement, however, is represented by a *negative* square: $-t^2$. This is what prevents time from being treated in exactly the same way as space, because as we all learned in school you cannot take the square root of a negative number. If you know x^2 then x has an easily understood meaning; the square root of 4, for example, is 2. But if you know $-t^2$, what does that tell us about t? What is the square root of, say, *minus* 9?

Hawking pointed out that the problem of the singularity at the beginning of the Universe – the 'edge' of time – can be resolved by taking on board an almost trivial mathematical device. Mathematicians know all about the square roots of negative numbers. They have been a standard feature of mathematics for more than 200 years, and mathematicians are able to manipulate them in their calculations with the aid of a single simple trick. They invented a 'number' called i, which is defined as 'the square root of minus one'. So $i \times i$ is equal to -1. If, now, you want to know the square root of -9, you say that -9 is equal to $(-1) \times 9$, and that the square root is equal to the square root of -1 multiplied by the square root of 9, which is simply $i \times 3$. Such 'imaginary numbers' can be manipulated in the same way as ordinary numbers – addition, multiplication, division and all the rest – and are an important part of many mathematical calculations. They provide a model for mathematicians to use in describing the undescribable, the world of square roots of negative numbers; and they operate by analogy with the way 'real' numbers operate.

Hawking's stroke of chutzpah was to suggest that our everyday understanding of time is wrong, and that a better model of the way the Universe works is obtained by changing over to using measurements in what he calls imaginary time, it. As far as the mathematics is concerned, this is a trivial change to make. It has about as much significance as a change in the choice of projection a map-maker uses in providing us with a picture of the Earth. For example, the traditional Mercator projection gives continents their correct shapes, by and large, but distorts their relative areas; while the Peters' projection, developed in the 1970s, shows continents in their correct relative proportions, but distorts their shapes. Both projections (and others) show the entire surface of the globe mapped on to a flat sheet of paper, and because it is impossible to represent the surface of a sphere perfectly on a flat sheet of paper no single projection can be said to be 'correct' while the others are 'incorrect'. They are just different.

In a similar way, mathematicians are free to choose many aspects of the coordinate systems they use in describing the positions of events in space and time. To take another geographical example, it is a historical accident that we choose to measure longitude relative to the meridian that passes through Greenwich, in London. Navigators could just as well use any of the other meridians, the imaginary lines joining the North and South poles of our planet, as 'longitude zero'.

Hawking's switch to 'imaginary time' is not quite that simple, but it involves only a change in choice of mathematical coordinates, and it has the dramatic effect of putting the time parameter in Einstein's equations on exactly the same footing as the space parameters. If time is measured in units of it, then when time measurements are squared we get units of $i^2 \times t^2$, which is simply $(-1) \times t^2$, or $-t^2$. Now, we have to multiply this negative number by the minus sign that comes into Einstein's equations themselves, which cancels out the (-1) we got from i^2 and leaves us with just t^2 (remember the old adage 'two negatives make a positive').

This change of model, or choice of a different mathematical analogy, has in effect made time, as far as Einstein's equations are concerned, *exactly* the same as space. And it turns out that this modest mathematical change removes the singularity from the equations.

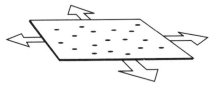

Figure 22
The expanding Universe can be thought of as like a rubber sheet being stretched simultaneously in all directions. The dots represent galaxies. Galaxies get further apart because the 'space' between them expands – not because they are moving through space.

The way we now have to think of the expanding Universe, says Hawking, is not in terms of a bubble of spacetime that appears out of a mathematical point (the singularity) and grows, but in terms of lines of latitude drawn on the surface of a sphere which stays a constant size. A tiny circle drawn around the north pole of the sphere represents the Universe when young – all of space is represented by the line that makes up the circle. As the Universe expands, it is represented by lines drawn further from the pole and closer to the equator, each one bigger than the previous circle. Moving from the pole towards the equator represents the 'flow' of time. Once past the equator, the 'universe' starts to shrink again as successive latitude circles get smaller, until they disappear at the south pole.

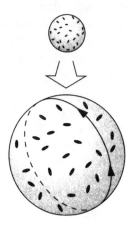

Figure 23
It may be that the Universe is just closed, although very nearly flat. In that case, it can be likened to the skin of an expanding soap bubble, with the dots representing galaxies, as in Figure 22. In this case, though, there is the curious possibility that you could travel right round the Universe and get back to where you started by always travelling in a straight line, just as it is possible to circumnavigate the Earth.

But what happens at the poles themselves – the beginning and end of time? There is no 'edge' to the sphere at these points, even though time is said to 'begin' at the north pole. Because time has been put on the same mathematical footing as space, the analogy with the geography of the Earth is perfect. At the North Pole of our planet, all directions are 'south', and there is no direction 'north' – but there is no edge to the planet there. At the north pole of Hawking's model of the Universe, all time directions are 'the future' and there is no direction of time corresponding to 'the past' – but there is no edge of time there. The singularity problem does not arise.

If you could travel backwards in time to the Big Bang itself, you would not disappear into a singularity, but would pass through the point (moment) of 'zero time' and find that you were heading off into the future again, in just the way that a person a little to the south of the North Pole of the Earth can walk due north, go across the pole and keep on walking in the same direction, but now finds that this direction is due south. The Universe is seen, on this picture, as a

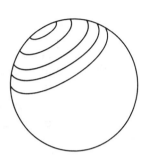

Figure 24
In Stephen Hawking's variation on the theme, both space and time (all four dimensions) are represented by the surface of a sphere. The Universe starts out as a tiny circle around the 'north pole' at time zero, and grows into successively larger circles, moving down towards the 'equator', as time passes. Then, it shrinks away to nothing as it moves through time towards the 'south pole'. But there is no 'edge' to spacetime, any more than there is an 'edge of the world' at the North Pole of the Earth. This representation is intended to show why it is meaningless to talk of a time 'before the Big Bang' or 'after the death of the Universe'.

completely self-contained package of spacetime and mass–energy, expanding out of nothing and contracting back into nothing.

All this has been achieved by a simple coordinate transformation, putting time on an equal footing with space. It's unfortunate that in mathematical jargon numbers involving i are traditionally called 'imaginary' numbers, because that means that Hawking's alternative time coordinate goes by the name 'imaginary time', which makes it sound like something out of a science fiction story, or Alice in Wonderland.[1] But this is, in fact, a mathematically respectable way of looking at things, which seems to be *more* physically reasonable than the traditional way of looking at things, since it does not contain the dreaded singularity.

There are other ways to explore the possibilities this raises. Hawking 'spatialized' time; Ilya Prigogine has said that his approach to a description of how things work is equivalent to 'temporalizing' space, treating creation as something that goes on everywhere in spacetime in some sense simultaneously. But I do not want to go into the details of that model here; all I want to do is to point out that Hawking's solution to the singularity problem is very much in the spirit of

---

[1] This choice of terminology is doubly unfortunate, because in fact what Hawking does is to treat time as if it were *imaginary space*; in the equations, $i\,t$ plays exactly the same role as x, y or z.

Augenstein's notion that anything in mathematics can be translated into a physically meaningful model of reality. Physics is work, in the same sense that carpentry is work, and it produces things out of raw materials. A carpenter makes furniture out of wood; a physicist makes models of the world out of mathematics. Who would have guessed, two centuries ago when the study of imaginary numbers was already a thriving branch of pure mathematics, that they would one day be applied to explain how the Universe came into being?

But the application, of course, had to wait until physicists and astronomers had developed a world view, or model, in which that problem was confronted in such a way that the solution in terms of imaginary time became apparent. So just how have physicists taken hold of the world and come up with their present description of reality?

GETTING A GRIP ON REALITY

One of the most recent, and most persuasive, explanations of how physicists actually set about finding (or inventing) their models of reality has been presented by Martin Krieger, of the University of Southern California, in his intriguing book *Doing Physics*. He has looked at the particular analogies and models developed during the second half of the twentieth century, and shown just how closely they are rooted in modern culture (in particular, for this period of time, the culture of the United States), and related to the analogies and models used by previous generations – the resemblance of QCD to QED and through that to Maxwell's equations being the most obvious example. There are echoes in some of this work of the way some philosophers – notably Karl Popper, from the 1930s onward[1] – have analysed the work of scientists in the twentieth century. But Krieger's background is in physics, which ought to make other physicists more willing to sit up and take notice, and his version of the story is both up to date and particularly persuasive.

Trained as a physicist, Krieger uses the terminology of physics in many respects, but translates it into everyday language. For example,

---

[1] See, for example, Popper, *The Logic of Scientific Discovery* (London: Hutchinson, 1959).

where a physicist might describe the properties of a system in terms of 'degrees of freedom', Krieger calls these properties 'handles' by which we can get hold of the system and get some idea of what it is like. A simple example would be the temperature of a box full of gas; this is a degree of freedom, and knowing the temperature tells us something about the overall conditions of the gas in the box. The position of an individual atom would be another example of a degree of freedom – but you don't need to know the position of every atom of gas in the box in order to know its temperature. Rather than trying to say what the world 'really' is, Krieger emphasizes that everything is based on analogy, and describes instead how physicists take hold of the world (using the handles provided by degrees of freedom) and describe it. The world may be 'like' many things – waves, or billiard balls, or whatever – without really *being* any of those things.

Krieger's use of analogy, though, extends much deeper than the examples I have used earlier. One of his insightful devices is to describe the workings of the subatomic world in terms of the workings of a factory, or of the economy of a country. An outsider, studying the way in which raw materials go into the factory and finished products emerge, cannot see the actual production process at work because it is hidden behind walls; but the observer can infer something about the production process by comparing the input and the output of the factory. The walls hide the details of the manufacturing process – they conceal degrees of freedom – and reduce the factory to a black box, for which the outsider can see that a certain input produces a certain output. This, says Krieger, is analogous to, for example, the way the electron cloud around an atom is responsible for its chemical behaviour, and conceals the inner workings of the atom. In chemical reactions, all that matters is the way the outermost electrons interact with the outermost electrons of other atoms, and you don't need to know anything about what holds the atoms themselves together.

The walls are important, because they simplify complex situations and make it possible to do worthwhile physics without knowing everything there is to know about a system. So physicists invent walls, the trick being to make sure that they are the right kind of walls. In effect, they conceal as many degrees of freedom as possible, and study the effect of changing the few degrees of freedom that remain – that is, the effect of taking hold of the system by its remaining 'handles' and shaking it.

Temperature is a good example. In many experiments involving gas in a box, the first thing physicists do is to wait until the temperature of the gas has settled down to some steady value – until it is in 'thermodynamic equilibrium'. Then, you no longer have to worry about temperature while you are studying some other properties of the gas for example, the way its pressure changes if you squeeze it into a box half the size (in practice, in order to carry out this simple experiment the box would have to be connected to a large object at a constant temperature – a 'heat bath' – to make sure that its own temperature did not change when the gas was squeezed). If you were to do the squeezing at the same time that the gas was being heated from outside, it would be much harder to disentangle all the changing degrees of freedom and get a picture of what was happening to the gas. If you choose the right degrees of freedom to study, the physics becomes straightforward; but if you make a mistake in your choice of degrees of freedom, the situation can become horribly complicated to unravel. As Steven Weinberg has commented, 'you may use any degrees of freedom you like to describe a physical system, but if you use the wrong ones, you'll be sorry'.[1]

Extending the factory analogy, Krieger likens the physicists' concept of particles to the individual workers in the factory, with their properties such as skills, mobility and wage demands. The properties of the 'workers' are described by the labels we attach to particles, labels we use to identify their charge, or their mass, or the strength of their response to the strong interaction. 'Particles,' he says, 'are designed to be localized and separated, stable and objective and named, and individual.'[2] The point, once again, is that physicists do not probe within the subatomic world and find particles; they start out with an idea of what billiard balls are like, and ask the kind of questions (choose the degrees of freedom) that invoke particle-like responses.

> We might wonder if we are misled by our everyday conception of billiard balls or of walls as we try to make Nature fit our naive intuitions. We might well be. But what is impressive is how we modify our naive intuitions, teaching ourselves to notice the right features of the everyday objects so that Nature is modelled by them.

A good example might be the quantum property known as spin. When physicists discovered that something more than just mass and charge

---

[1] Quoted in Krieger, *Doing Physics*, p. 30.
[2] This and the next quote from Krieger, *Doing Physics*, pp. 22–3.

was needed as a label for the electron, they drew an analogy with the way a billiard ball rotates, to find a new label. The analogy is not exact, because it turns out that, if you insist on treating an electron as a spinning particle, it must be envisaged as rotating through not 360 degrees but 720 degrees (*two* complete revolutions) to get back to where it started.[1] But physicists have trained themselves to think of this bizarre property as an analogue of the spin of a billiard ball, or of the rotation of the Earth.

The third component of the physicists' world, alongside walls and workers, is the field. A field is the very antithesis of a particle – spread out instead of localized, smoothly changing instead of being sharp-edged. But fields are always connected to particles, and as Krieger points out a 'perfect' particle would be completely self-contained and have no handles by which we might shake it – it is only because gravitational field, or electromagnetic field, or whatever, leaks out of particles that we know they are there at all.

But this still does not mean that the field is 'real', any more than the particles are real, or that an electron really spins on its axis like a top. Or rather, as I prefer to see it, *all* the models are real, even if they are incomplete. As Krieger argues, what other reality is there, apart from the models? Like Pickering, Krieger discusses the way physicists learn their trade and make progress by imitating the techniques that have proved successful in the past, one of the most powerful of which has been to assume that everything is made of smaller parts. He discusses the power of the clockwork analogy, and points out (p. 33) that 'a clock does many fewer (but perhaps more interesting) things than do all of its components separately' – another example of the way restricting the number of degrees of freedom may be a good thing. But he doesn't detail the way in which Maxwell arrived at his famous wave equations through the intermediate step of a clockwork-like system of interacting cogs and gears.

Traditionally, this step is regarded as a dispensable device, like a crutch which can be discarded once a patient has learned to walk without it. But the fact is that it worked. It may be tedious and unappealing, but it does provide a model for the way electromagnetic forces are transmitted. Field theory is 'better' because it seems simpler

---

[1] Richard Feynman has given a delicious example of how two revolutions can get you back where you started, using a cup of tea as his model, on p. 29 of *Elementary Particles and the Laws of Physics.*

and more straightforward to us; but the fact that a clockwork model can be made to work, even if it seems ugly and crude to us, is an important reminder that the analogies we like best need not necessarily represent a unique truth about the way the world works. When physicists say that Nature does work in a certain way, Krieger argues, what they are really saying is that the models can be legitimately *made* to work in that way.

Here's another example of a largely discarded, but still viable, metaphor. When I discussed the creation of electron–positron pairs out of pure energy, I did so in terms of energy being converted into mass, in line with $E = mc^2$. But when Paul Dirac first raised the possibility of the existence of what are now known as antiparticles, at the end of the 1920s, he came up with a different model. In this version of reality, the 'nothingness' of the vacuum is filled with a sea of electrons, with every possible *negative* energy level occupied. We do not notice these electrons, because they are everywhere, and provide no basis for discriminating each other from their surroundings. If a wall is painted a uniform colour (say, red), then each point on the wall is as red as any other, and no point stands out. An ordinary (positive-energy) electron is 'noticed' because it is different from whatever is next door to it, like a spot of blue paint on the red background.

On this picture, the creation of an electron–positron pair happens when a sufficiently energetic photon strikes one of the negative-energy electrons and gives it enough energy to 'promote' it into a positive-energy state. It becomes a 'real' electron in the everyday world (a blue spot), and leaves a hole behind in the sea of negative-energy electrons (a white spot against the red background). This hole has all the properties of an electron with positive charge – a positron. For example, if there is a positive electric charge nearby, all of the negative-energy electrons will try to move towards that charge. Where they are jammed shoulder to shoulder, they cannot move. But the electron next to the hole can shuffle forward by jumping into the hole, leaving a gap behind, and so on down the line. The effect is that the hole moves away from the positive charge – it is repelled, just as a positively charged particle would be. In the negative-energy sea, the *absence* of an electron provides a distinction with its surroundings, the 'sharp edge' that is one of the distinguishing characteristics of a particle. The hole remains, and behaves like a particle, until a positive-energy electron falls into the hole, giving up energy in the form of electromagnetic radiation as it does so.

Like Maxwell's cogwheels and vortices, this model of particle–antiparticle interactions is now seen as an intermediate step on the way to the 'true' picture of creation of particles out of pure energy. But it is an entirely reasonable, self-consistent model, which can be used as the basis for calculations and which can correctly predict the properties of positrons as measured in experiments – and, remember, there is yet another equally satisfactory model which explains positrons as electrons moving backwards through time. It may make us uncomfortable to think of the Universe as being full of negative-energy electrons, but that is our problem, not the Universe's. We are free to choose the degrees of freedom we want to investigate, and these choices determine the properties we ascribe to Nature. Analogy is *everything* in physics, and as long as the models we construct are self-consistent and make predictions that can be tested and confirmed by experiments then we are free to use any analogies, and choose any degrees of freedom, that we wish. Which brings me back to the question of which of the many quantum interpretations, if any, can be regarded as a 'best buy'.

THE BULK-BUYING APPROACH TO QUANTUM REALITY

The best answer, it seems to me, might well be to buy the lot. Each of the interpretations is a viable model, and each of them provides us with useful insights into the way the world works. Indeed, it is quite reasonable to regard each of the quantum interpretations as a degree of freedom in its own right, and by applying Weinberg's dictum we are free to choose the interpretation that most suits our needs in any particular situation. Choose the wrong one, and you'll be sorry – for example, if you choose the Copenhagen Interpretation to 'explain' what happens to Schrödinger's cat. But choose the right one – in this case, the many-worlds interpretation – and everything is straightforward. A good physicist should carry every quantum interpretation is his or her toolkit, and should apply the right one for the job in hand when confronted with a particular quantum puzzle.

To prove the point, here is a quick reminder of some of the interpretations on offer, and how they relate to Bell's theorem, the most important development in quantum physics in the second half of the twentieth century. Any acceptable version of quantum reality must be compatible with the results of the Aspect experiment – and they *all* are!

The good old Copenhagen Interpretation has no difficulty with Bell's theorem and the Aspect experiment, because Niels Bohr and his colleagues told us all along that the outcome of an experiment depends on the entire experimental set-up. If both slits are open in the experiment with two holes, we get interference; if only one slit is open, we do not. And if the entire experimental set-up includes photons on opposite sides of the Galaxy, then we have to take account of both photons even if that does involve 'spooky action at a distance'. Equally, if reality is created by the act of making a measurement, all we have to do to understand the results of the Aspect experiment in terms of this interpretation is to accept that the reality that is created is not necessarily just the reality in the immediate locality where the measurement is being made, but also the reality far away, in places which light signals from the measurement have not yet had time to reach.

Alternatively, the world may be 'really real', in the way that David Bohm and his followers suggest. But if so, it must be, according to Bohm, in a state of undivided wholeness so that, again, a prod in one place can produce a response far away, non-locally and immediately. In both this and the related idea of real particles with real properties that are influenced by a pilot wave that obeys the statistical laws, the instantaneous 'communication' affects the outcome of experiments by taking account of the state of the rest of the Universe, but still manages not to allow any transmission of faster-than-light signals containing useful information between human observers.

The many-worlds interpretation is in a slightly different category, because it allows all possible outcomes to all possible experiments to be equally real. But, as I have mentioned, it is certainly non-local, since the choice of outcomes of a quantum event here on Earth is causing the creation of multiple copies of reality in far distant galaxies instantaneously (and changes in those galaxies are instantly causing reality here on Earth to split into multiple copies). But still, it does work, as a self-consistent interpretation of quantum reality.

John Bell, discussing the rival interpretations of quantum theory, puts things in perspective:

To what extent are these possible worlds fictions? They are like literary fiction in that they are free inventions of the human mind. In theoretical physics sometimes the inventor knows from the beginning that the work is fiction, for example when it deals with a simplified world in which space has only one or two dimensions instead of three. More often it is not known till later, when the hypothesis has proved wrong, that fiction is involved. When being serious, when not exploring deliberately simplified models, the theoretical physicist differs from the novelist in thinking that maybe the story might be true.[1]

But such hopes are ill-founded. *All* models are deliberately simplified, by our choice of which degrees of freedom to use as handles on reality; and all models of the world beyond the reach of our immediate senses are fictions, free inventions of the human mind. You are free to choose whichever of the quantum interpretations most appeals to you, or to reject all of them, or to purchase the entire package and use a different interpretation according to convenience, or the day of the week, or whim. Reality is in very large measure what you want it to be.

Still, though, almost everybody wants to know 'the answer'. The quest for a *really* real model is what drives theoretical physicists, just as it motivates other folk to study philosophy or to subscribe to a particular religion. I still have this hankering myself, even though the logical part of my mind tells me that the search is fruitless, and that all we can ever hope to find is a self-consistent myth for our times. So, in spite of everything, I do not intend to leave you without offering what I consider to be the best buy for now in the quantum-reality marketplace, an interpretation which not only brings clearly to the forefront the whole business of non-locality, but which also provides a set of analogies and metaphors which, I believe, are about to transform the way physicists think about the world.

In *Doing Physics*, Martin Krieger mentions many analogies which prove useful in understanding what physicists do. The factory and its workers, economies, the familiar clockwork models, and even kinship systems, have their place in his discussion. But, he says (p. xix), 'other major analogies, such as that of evolution and organism, seem to play a much smaller role in most of physics'.

I believe that this is an historical oversight which is now being rectified. As I have discussed in my book *In the Beginning*, by treating objects such as galaxies, and even the Universe itself, as if they were

---

[1] Bell, *Speakable and Unspeakable*, pp. 194–5.

living, evolving organisms, astronomers and cosmologists are gaining new insights into the nature of the world, its origins and its ultimate fate. And key concepts related to the way living things operate also turn up in my favourite quantum fiction, the so-called transactional interpretation. I do not claim that it is anything more than a fiction; all scientific models are simply Kiplingesque 'just so' stories that give us a feeling that we understand what is going on, without necessarily incorporating any ultimate answers about the Universe. But if you want a story you can believe in for the time being, and which will probably last a long while before it is replaced by something better (or simply more fashionable), then the transactional interpretation is the one I recommend. The time has come for me to nail my colours to the mast, as we rejoin all those readers who have skipped everything since the Prologue, and to present a version of reality which really does take away all the mystery from the quantum mysteries.

EPILOGUE

The Solution – A Myth for our Times

The central problem that we have to explain, in order to persuade ourselves that we understand the mysteries of the quantum world, is encapsulated in the story of Schrödinger's kittens that I told in the Prologue. The experiment is set up, remember, in such a way that the two kittens have been separated far apart in space, but are each under the influence of a 50:50 probability wave, associated with the collapse of an electron wave function to become a 'real' particle in just one or other of their two spacecraft. At the moment when one of the capsules is opened and an intelligent observer notices whether or not the electron is inside, the probability wave collapses and the fate of the kitten is determined – and not just the fate of the kitten in that capsule, but also, *simultaneously*, that of the other kitten in the other capsule, on the other side of the Universe.

At least, that is the standard Copenhagen Interpretation version of the correlation between the two kittens, and whichever quantum interpretation you favour, the Aspect experiment and Bell's inequality show that once quantum entities are entangled in an interaction then they really do behave, ever afterwards, as if they are parts of a single system under the influence of Einstein's 'spooky action at a distance'. The whole is greater than the sum of its parts, and the parts of the whole are interconnected by feedbacks – feedbacks which seem to operate instantaneously.

This is where we can begin to make a fruitful analogy with living systems. A living system, such as your own body, is certainly greater than the sum of its parts. A human body is made up of millions of cells, but it can do things that a heap of the appropriate number of cells could never do; the cells themselves are alive in their own right, and they can do things that a simple chemical mixture of the elements they contain could not do. In both cases, one of the key reasons why the living cells and living bodies can do such interesting things is that there are feedbacks which convey information – from one side of the cell to another, and from one part of the body to another. At a deep level, inside the cells these feedbacks may involve chemical messengers which convey raw materials to the right places and use them to construct complex molecules of life. At a gross human level, just about every routine action, such as the way my fingers are moving to strike the right keys on my computer keyboard to create this sentence, involves feedbacks in which the brain constantly takes in information from senses such as sight and touch and uses that information to modify the behaviour of the body (in this case, to determine where my fingers will move to next).

This really is feedback, a two-way process, not simply an instruction from the brain to tell the fingers where to go. The whole system is involved in assessing where those fingers are now, and how fast (and in what direction) they are moving, checking that the pressure on the keys is just right, going back (very often, in my case!) to correct mistakes, and so on. Even a touch-typist is constantly adjusting the exact movements of the fingers in response to such feedbacks, in the same way that you can ride a bicycle by constantly making automatic adjustments in your balance to keep yourself upright. If you knew nothing about those feedbacks, and had no idea that the different parts of the body were interconnected by a communications system, it would seem miraculous that the elongated lumps of flesh and bone on the ends of my hands could 'create' an intelligent message by poking away at the keyboard – just as it seems miraculous, unless we invoke some form of communication and feedback, that the polarization states of two photons flying out on opposite sides of an atom can be correlated in the way that the Aspect experiment reveals. The one big difference, the hurdle that we have to overcome, is the *instantaneous* nature of the feedback in the quantum world. But that is explained by the nature of light itself, both in the context of relativity theory and from the right perspective on the quantum nature of electrodynamics. That perspective is the so far relatively unsung Wheeler–Feynman model of electromagnetic radiation – a model which can also provide striking insights into the way gravity works.

MAKING THE MOST OF MASS

Feynman's unsung insight suggested, more than half a century ago, that the behaviour of electromagnetic radiation, and the way in which it interacts with charged particles, could be explained by taking seriously the fact that there are two sets of solutions to Maxwell's equations, the equations that describe electromagnetic waves moving through space like ripples moving across the surface of a pond. One set of solutions, the 'common-sense' solutions, describes waves moving outward from an accelerated charged particle and forwards in time, like ripples spreading from the point where a stone has been dropped into the pond. The second set of solutions, largely ignored even today, describes waves travelling backwards in time and converging onto charged particles, like ripples that *start* from the edge of the pond and converge onto a point in the middle of the pond. As I discussed in Chapter Two, when proper allowance is made for both sets of waves interacting with all the charged particles in the Universe, most of the complexity cancels out, leaving only the familiar common-sense (or 'retarded') waves to carry electromagnetic influences from one charged particle to another. But as a result of all these interactions, each individual charged particle – including each electron – is *instantaneously* aware of its position in relation to all the other charged particles in the Universe. The one tangible influence of the waves that travel backwards in time (the 'advanced' waves) is that they provide feedback which makes every charged particle an integrated part of the whole electromagnetic web. Poke an electron in a laboratory here on Earth, and in principle every charged particle in, say, the Andromeda galaxy, more than two million light years away, *immediately* knows what has happened, even though any retarded wave produced by poking the electron here on Earth will take more than two million years to reach the Andromeda galaxy.

Even supporters of the Wheeler–Feynman absorber theory usually stop short of expressing it that way. The conventional version (if anything about the theory can be said to be conventional) says that our electron here on Earth 'knows where it is' in relation to the charged particles everywhere else, including those in the Andromeda galaxy. But it is at the very heart of the nature of feedback that it works both ways. If *our* electron knows where the Andromeda galaxy is, then for sure the Andromeda galaxy knows where our electron is.

The result of the feedback – the result of the fact that our electron has to be considered not in isolation but as part of a holistic electromagnetic web filling the Universe – is that the electron resists our attempts to push it around, because of the influence of all those charged particles in distant galaxies, even though no information-carrying signal can travel between the galaxies faster than light.

Now this explanation of why charged particles experience radiation resistance is rather similar to another puzzle, mentioned earlier, that has long plagued physicists. Why do ordinary lumps of matter resist being pushed around, and how do they know how much resistance to offer when they are pushed? Where does inertia itself come from?

Galileo seems to have been the first person to realize that it is not the velocity with which an object moves but its acceleration which reveals the effect of forces acting upon it. On Earth, friction – one of those external forces – is always present, and slows down (decelerates) any moving object, unless you keep pushing it. But without the influence of friction objects would keep moving in straight lines forever, unless they were pushed or pulled by forces.

This became one of the cornerstones of Newton's laws of mechanics. Things move at constant velocity through empty space (relative to some absolute standard of rest), he argued, unless accelerated by external forces. For an object with a given mass, the acceleration produced by a particular force is given by dividing the force by the mass.

One intriguing aspect of this discovery is that the mass which comes into the calculation is the same as the mass involved in gravity. It isn't immediately obvious that this should be so. Gravitational mass determines the strength of the force which an object extends out into the Universe to tug on other objects; inertial mass, as it is called, determines the strength of the response of an object to being pushed and pulled by outside forces – not just gravity, but *any* outside forces. And they are the same. The 'amount of matter' in an object determines both its influence on the outside world, and its response to the outside world.[1] This already looks like a feedback at work, a two-way process linking each object to the Universe at large. But until very recently,

---

[1] Don't be confused by the fact that an object weighs less on the Moon than it does on Earth; this is not because the object itself changes, but because the gravitational force at the surface of the Moon is less than the gravitational force at the surface of the Earth. It is the outside force that is less on the Moon, and the inertial response of the object matches that reduced outside force, so that it 'weighs less'.

nobody had any clear idea how the feedback could work.

Newton himself described a neat experiment which seems to show that there really is a preferred frame of reference in the Universe, and later philosophers said that this experiment indicates just what it is that defines the absolute standard of rest. Writing in the *Principia* in 1686, Newton described what happens if you take a bucket of water hung from a long cord, twist the cord up tightly, and then let go. The bucket, of course, starts to spin as the cord untwists. At first, the surface of the water in the bucket stays level, but as friction gradually transfers the spinning of the bucket to the water itself, the water begins to rotate as well, and its surface takes up a concave shape, as 'centrifugal force' pushes water out to the sides of the bucket. Now, if you grab the bucket to stop it spinning, the water carries on rotating, with a concave surface, but gradually slows down, becoming flatter and flatter, until it stops moving and has a completely flat surface.

Newton pointed out that the concave shape of the surface of the rotating water shows that it 'knows' that it is rotating. But what is it rotating relative to? The relative motion of the bucket and water seems completely unimportant. If the bucket and the water are both still, with no relative motion, the water is flat; if the bucket is rotating and the water is not, the surface is still flat even though there is relative motion between the water and the bucket; if the water is rotating and the bucket is not, there is relative motion between the two and the surface is concave; but if the water and the bucket are both rotating, so that once again there is no relative motion between the water and the bucket, the surface is concave. So, Newton reasoned, the water 'knows' whether or not it is rotating relative to absolute space.

In the eighteenth century, the philosopher George Berkeley offered another explanation. He argued that all motion must be measured relative to something tangible, and he pointed out that what seems to be important in the famous bucket experiment is how the water is moving relative to the most distant objects known at the time, the fixed stars. We now know, of course, that the stars are relatively near neighbours of ours in the cosmos, and that beyond the Milky Way there are many millions of other galaxies. But Berkeley's insight still holds. The surface of a bucket of water will be flat if the water is not rotating relative to the distant galaxies, and it will be curved if the water is rotating relative to the distant galaxies. And acceleration seems also to be measured relative to the distant galaxies – that is, relative to the average distribution of all the matter in the Universe. It is as if, when you try to push something around, it takes stock of its situation relative to all the matter in the Universe, and responds accordingly. It is somehow held in place by gravity, which is why gravitational and inertial mass are the same.

This idea that inertia is indeed produced by the response of a material object to the Universe at large is often known as Mach's Principle, after the nineteenth-century Austrian physicist Ernst Mach, whose name is immortalized in the number used to measure speeds relative to the speed of sound, but who also thought long and hard about the nature of inertia.

As I have mentioned, Mach's ideas, essentially an extension of those of Berkeley, strongly influenced Einstein, who argued that the identity between gravitational and inertial mass does indeed arise because inertial forces are really gravitational in origin, and tried to incorporate Mach's Principle – the feedback of the entire Universe on any gravitational mass – into his general theory of relativity. It is fairly easy to make a naive argument along these lines. All the mass in all the distant galaxies (and anything else) reaches out with a gravitational influence to hold on to everything here on Earth (and everywhere else), including, say, the pile of computer disks sitting on my desk. When I try to move one of those disks, the amount of effort I have to put into the task is a measure of how strongly the Universe holds that disk in its grip.

But it is much harder to put all this on a secure scientific footing. How does the disk 'know', instantaneously, just how much it should resist my efforts to move it? One appealing possibility (in the naive picture) is that by poking at an object and changing its motion we make it send some sort of gravitational ripple out into the Universe, and that this ripple disturbs everything else in the Universe, so that a kind of echo comes back, focusing down on the disturbed object and trying to maintain the status quo. But if signals, including gravitational ripples, can only travel at the speed of light, it looks as if it might take just about forever for the echo to get back and for the disk to decide just how it ought to respond to being pushed around.

Unless, of course, there is some way of incorporating the principle of the time-symmetric Wheeler–Feynman absorber theory into a description of gravity, so that some of the gravitational ripples involved in this feedback travel backwards in time. But since the Wheeler–Feynman theory of electromagnetic radiation came some 30 years after Einstein's theory of gravity, and nobody took it very seriously even then, this resolution of the puzzle posed by Mach's Principle has had to wait a long time to be put on a proper mathematical footing.

Ever since Einstein came up with his general theory, there has been argument about whether or not it does incorporate Mach's Principle in a satisfactory way. It does at least go some way towards including Mach's Principle, because the behaviour of an object at any location in space depends on the curvature of spacetime at that location, which is determined by the combined gravitational influence of all the matter in the Universe. But it still seems to beg the question of how quickly the 'signals' that determine the curvature of spacetime get from one place to another. Since those distant galaxies are themselves moving, their influence ought to be constantly changing. Do these changes propagate only at the speed of light, or instantaneously? And if instantaneously, how?

One intriguing aspect of the debate is that Einstein's equations only produce anything like the right kind of Machian influences if there is enough matter in the Universe to bend spacetime back on itself gravitationally. In an 'open' Universe, extending to infinity in all directions, the equations can never be made to balance with a finite amount of inertia. This used to be an argument against claiming that the general theory incorporates Mach's Principle, because people thought that the Universe was 'open'; but as we saw in Chapter Two all that has changed, and there now seems to be compelling evidence that the Universe is indeed 'closed'. Which, of course, is one reason why the Wheeler–Feynman absorber theory itself is now taken more seriously.

In 1993, Shu-Yuan Chu, of the University of California, published a paper which shows which way the wind is now blowing.[1] Chu had been looking at Bell's inequality in the context of a variation on Wheeler–Feynman theory, so I wrote to ask what else he had been working on. Among other things, that turned out to include an investigation of how to do quantum mechanics in the presence of gravity, which neatly combines some of the latest ideas in particle physics with a time-symmetric Wheeler–Feynman model to show where gravity itself comes from, with inertia explained along the way. At the time of writing (March 1994), this work exists only as a University of California 'preprint', number UCR-HEP-T117; describing it is as close as any book of this kind can ever get to offering you

[1] *Physical Review Letters* 71 (1993), p. 2847.

a glimpse of current research, and it wraps up so many ideas in such a neat package that it cannot be allowed to pass without mention.

STRINGING GRAVITY TOGETHER

First, I need to take you on a tiny detour to explain the particle-physics end of the story. In the 1990s, particle physicists no longer halt their journey into the innermost recesses of matter at the level of particles like electrons and quarks. In yet another recapitulation of the historical processes of taking 'fundamental' particles apart to find out what is inside them, in the middle of the 1980s some particle physicists became intrigued by the discovery that the properties of particles like quarks and electrons could be explained if they were composed of

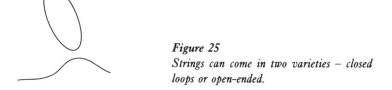

Figure 25
Strings can come in two varieties – closed loops or open-ended.

much smaller entities called strings. As their name suggests, these 'new' entities differ from the familiar billiard-ball models of other particles by having length – an extension in one dimension, literally like a tiny piece of string.

But 'tiny' is the operative word. A typical string would be only 10^{-35} of a metre long, so that it would take 10^{20} such strings, laid end to end, to stretch across the diameter of a proton. There is no direct experimental evidence that such strings exist. Experiments to probe interactions on such a scale would require much more energy than any conceivable particle accelerator that could be built on Earth could provide. But the possibility of their existence is based on a well-founded theory of the way the interactions of the particle world work, derived in part from the archetypal QED and QCD approaches to a theory of everything.

Now, I have argued that none of our theories and models provide 'the truth' about the particle world, and that all of them are more or less successful attempts to provide a picture we can understand and models we can use to make predictions with. On that basis, string theory is very successful indeed. Although nobody has seen a string, or even detected one in a particle-accelerator experiment, properties such as charge can be explained as being 'tied' to the ends of the strings, and particle interactions can be explained in terms of strings colliding and joining, or splitting apart. It even turns out that closed loops of vibrating string, like tiny elastic bands, automatically have the properties needed to act as gravitons – the particles that carry the gravitational force, equivalent to the way photons carry the electromagnetic force. The whole package is self-consistent, logical, and (for the mathematically adept) as good an explanation of how the world works as any other. The one drawback is that there is, as yet, no way to apply Newton's ultimate experimental test. But that hasn't stopped theorists trying to use the theory to explain already known features of the Universe – which is just what Chu has done.

His investigation of gravity is part of a larger attempt to explain interactions at this level with the aid of time-symmetric descriptions based on the Wheeler–Feynman approach. This process eliminates the idea of 'fields' (for example, electromagnetic and gravitational fields) as independent entities. Particles interact with one another in a time-symmetric way, exchanging advanced and retarded 'messages' in a continuous feedback, and what we are used to thinking of as a continuous field, such as gravity, is built up by averaging over all the interactions involving little pieces of matter. The continuous gravitational field emerges from this averaging process on a scale which has to be large compared with the scale of the particles involved – but if the particles are actually pieces of string so small that it takes 10^{20} of them to stretch across a proton, that means that even on the scale of a proton gravity will seem to be very smooth and continuous. 'The curvature of space-time,' says Chu, 'is just a reflection of the patterns of the tapestry of motion woven with the world-sheets of the strings.'

One of the implications of this approach is that the description of particle motion in terms of Newtonian, classical trajectories emerges from a kind of statistical averaging over the behaviour of particles. 'The strings execute small scale jitters about their particle-like paths ... after we average over the strong jitters.' This has echoes both of Feynman's path-integral (sum-over-histories) approach and of Ilya

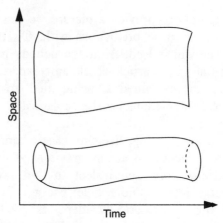

Figure 26
When open-ended strings move through spacetime, they sweep out 'world sheets'; when closed loops move through spacetime, they sweep out 'world tubes'.

Prigogine's statistical approach to understanding the particle world, which develops from thermodynamics. This is not the place to go into all the details – that would require another book as big as this one – but both Prigogine and Chu have constructed descriptions of reality in which the statistics come first, and the classical particle trajectories emerge from the statistics. Both in the classical and the quantum worlds, in Chu's words, 'the foundation of mechanics appears to be built on statistics ... one should derive mechanics from statistics and not the other way around'.

The link with thermodynamics is explicit. The key concept in thermodynamics is entropy, a property which measures how close a system is to equilibrium. Chu's description shows that Einstein's

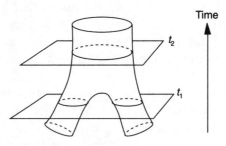

Figure 27
Two loops of string moving through spacetime and merging together create a pair of spacetime trousers.

equation of motion is the correct description of particle trajectories under the equilibrium condition of maximum entropy. But, as in the original Wheeler–Feynman theory (and in attempts to incorporate Mach's Principle into the general theory of relativity) there must be complete absorption of all the radiation from strings today into the future – in other words, the Universe must be closed. In a neat summary in a letter to me at the end of 1993, Chu spelled out his conclusions: 'Classical mechanics describes the equilibrium condition (hence the absence of any probabilistic statements in classical mechanics); quantum mechanics describes the fluctuations; and the path integral formalism follows from summing over the huge number of strings in the system.'

There is even a bonus, for anyone who has followed the cosmological debate in recent years. Einstein's description of the Universe, the equations of the general theory, includes a number, dubbed the cosmological constant, which has been an embarrassment to astronomers for more than 70 years. There is no way to predict the value of this number from Einstein's equations, and it seems as if it could have any value. Yet observations of the way the Universe at large is expanding suggest that it must be very, very close to zero. Even a small cosmological constant would have a profound influence on the way the Universe is seen to be expanding. Chu's description of gravity, however, becomes exactly the same as Einstein's description over distances much larger than the length of a piece of string, but with no cosmological constant term at all.

Getting back to Bell's inequality, the problem is that the experiments show that there are instantaneous correlations between separated particles. But as Chu points out in that *Physical Review Letters* paper: 'Instantaneous correlation between two spatially separated particles can be established through a third particle, which correlates with one of the two particles through the advanced interaction and correlates with the other particle through the retarded interaction.'

This was his motivation to try to incorporate the Wheeler–Feynman approach into a description of quantum mechanics, and then into a description of gravity, using string theory. What he didn't realize at the time was that the philosophical foundations for such an approach had already been laid by John Cramer, of the University of Washington, Seattle, in a series of largely unsung papers published in the 1980s. Cramer's 'transactional interpretation' of quantum mechanics uses exactly this approach, and the success of Chu's application of similar

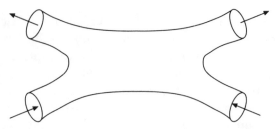

Figure 28
In the context of string theory, an interaction between two particles is reinterpreted in terms of world tubes merging and breaking apart. This kind of diagram can be made much more complex, involving many loops of string, in the same way that complications arise in the calculation of the magnetic moment of an electron (see Figure 13).

ideas to string theory and gravity suggests strongly that this is going to be a fertile field of physics in the immediate future. As Chu said when I told him about Cramer's work, 'Had I known advanced interactions have already been accepted as a possibility in these discussions, it would certainly have lessened my anxiety in pursuing the string-theory generalization of Wheeler–Feynman's time-symmetric electrodynamics.'

Well, prepare to cast all such anxieties aside – for here it is, the interpretation that provides the best all-round picture of how the world works at the quantum level, for anyone who wants to have a single 'answer' to the puzzles posed by Bell's inequality, the Aspect experiment, and the fate of Schrödinger's kittens.

THE SIMPLE FACE OF COMPLEXITY

The original version of the Wheeler–Feynman theory was, strictly speaking, a classical theory, because it did not take account of quantum processes. Nevertheless, by the 1960s researchers had found that there are indeed only two stable situations that result from the complexity of overlapping and interacting waves, some going forwards in time and some backwards in time. Such a system must end up dominated either by retarded radiation (like our Universe) or by advanced radiation (equivalent to a universe in which time ran backward). In the early

1970s, a few cosmologists, intrigued by the puzzle of why there should be an arrow of time in the Universe at all, developed variations on the Wheeler–Feynman theory that did take on board quantum mechanics. In effect, they developed Wheeler–Feynman versions of QED. Fred Hoyle and Jayant Narlikar used a path-integral technique, while Paul Davies used an alternative mathematical approach called S-matrix theory. The details of the mathematics do not matter; what does matter is that in each case they found that Wheeler–Feynman absorber theory can be turned into a fully quantum-mechanical model.

The reason for the interest of cosmologists in all this is the suggestion – still no more than a suggestion – that the reason why our Universe should be dominated by retarded waves, and that there should, therefore, be a definite arrow of time, is connected with the fact that the Universe itself shows time-asymmetry, with a Big Bang in the past and (probably) ultimate collapse into a Big Crunch in the future. Wheeler–Feynman theory provides a way for particles here and now to 'know' about the past and future states of the Universe – these 'boundary conditions' could be what selects out the retarded waves for domination.

But all of this still applied only to electromagnetic radiation. The giant leap taken by John Cramer was to extend these ideas to the wave equations of quantum mechanics – the Schrödinger equation itself, and the equations describing the probability waves, which travel, like photons, at the speed of light. His results appeared in an exhaustive review article published in 1986,[1] but made so little impact that, for example, when Chu was developing his ideas based on string theory in 1993 he had never heard of Cramer's interpretation.

In order to apply the absorber-theory ideas to quantum mechanics, you need an equation, like Maxwell's equations, which yields two solutions, one equivalent to a positive energy wave flowing into the future, and the other describing a negative energy wave flowing into the past. At first sight, Schrödinger's famous wave equation doesn't fit the bill, because it only describes a flow in one direction, which (of course) we interpret as from past to future. But as all physicists learn at university (and most promptly forget) the most widely used version of this equation is incomplete. As the quantum pioneers themselves

---

[1] 'The transactional interpretation of quantum mechanics', *Reviews of Modern Physics* **58** (1986), p. 647.

realized, it does not take account of the requirements of relativity theory. In most cases, this doesn't matter, which is why physics students, and even most practising quantum mechanics, happily use the simple version of the equation. But the full version of the wave equation, making proper allowance for relativistic effects, is much more like Maxwell's equations. In particular, it has two sets of solutions – one corresponding to the familiar simple Schrödinger equation, and the other to a kind of mirror-image Schrödinger equation describing the flow of negative energy into the past.

This duality shows up most clearly in the calculation of probabilities in the context of quantum mechanics. The properties of a quantum system are described by a mathematical expression, sometimes known as the 'state vector' (essentially another term for the wave function), which contains information about the state of a quantum entity – the position, momentum, energy and other properties of the system (which might, for example, simply be an electron wave packet). In general, this state vector includes a mixture of both ordinary ('real') numbers and imaginary numbers (those numbers involving i, the square root of -1). Such a mixture is called a complex variable, for obvious reasons; it is written down as a real part plus (or minus) an imaginary part. The probability calculations needed to work out the chance of finding an electron (say) in a particular place at a particular time actually depend on calculating the square of the state vector corresponding to that particular state of the electron. But calculating the square of a complex variable does not simply mean multiplying it by itself. Instead, you have to make another variable, a mirror-image version called the complex conjugate, by changing the sign in front of the imaginary part: if it was $+$ it becomes $-$, and vice versa. The two complex numbers are then multiplied together to give the probability. But for equations that describe how a system changes as time passes, this process of changing the sign of the imaginary part and finding the complex conjugate is equivalent to reversing the direction of time! The basic probability equation, developed by Max Born back in 1926, itself contains an explicit reference to the nature of time, and to the possibility of two kinds of Schrödinger equations, one describing advanced waves and the other representing retarded waves. It should be no surprise, after all this, to learn that the two sets of solutions to the fully relativistic version of the wave equation of quantum mechanics are indeed exactly these complex conjugates. But in time-honoured tradition, for some 70 years most physicists have largely ignored one of the two sets of solutions because 'obviously' it didn't make sense to talk about waves travelling backwards in time!

The remarkable implication is that ever since 1926, every time a physicist has taken the complex conjugate of the simple Schrödinger equation and combined it with this equation to calculate a quantum probability, he or she has actually been taking account of the advanced-wave solution to the equations, and the influence of waves that travel backwards in time, without knowing it. There is no problem at all with the mathematics of Cramer's interpretation of quantum mechanics, because the mathematics, right down to Schrödinger's equation, is *exactly the same* as in the standard Copenhagen Interpretation. The difference is, literally, only in the interpretation. As Cramer put it in that 1986 paper (p. 660), 'the field in effect becomes a mathematical convenience for describing action-at-a-distance processes'. This is exactly the view Chu arrived at independently seven years later. So, having (I hope) convinced you that this approach makes sense, let's look at how it explains away some of the puzzles and paradoxes of the quantum world.

SHAKING HANDS WITH THE UNIVERSE

The way Cramer describes a typical quantum 'transaction' is in terms of a particle 'shaking hands' with another particle somewhere else in space and time. You can think of this in terms of an electron emitting electromagnetic radiation which is absorbed by another electron, although the description works just as well for the state vector of a quantum entity which starts out in one state and ends up in another state as a result of an interaction – for example, the state vector of a particle emitted from a source on one side of the experiment with two holes and absorbed by a detector on the other side of the experiment. One of the difficulties with any such description in ordinary language is how to treat interactions that are going both ways in time simultaneously, and are therefore occurring instantaneously as far as clocks in the everyday world are concerned. Cramer does this by effectively standing outside of time, and using the semantic device of a description in terms of some kind of pseudotime. This is no more than a semantic device – but it certainly helps to get the picture straight.

It works like this. When an electron vibrates, on this picture, it attempts to radiate by producing a field which is a time-symmetric mixture of a retarded wave propagating into the future and an advanced wave propagating into the past. As a first step in getting a picture of what happens, ignore the advanced wave and follow the story of the retarded wave. This heads off into the future until it encounters an electron which can absorb the energy being carried by the field. The process of absorption involves making the electron that is doing the absorbing vibrate, and this vibration produces a new retarded field which exactly cancels out the first retarded field. So in the future of the absorber, the net effect is that there is no retarded field.

But the absorber also produces a negative-energy advanced wave travelling backwards in time to the emitter, down the track of the original retarded wave. At the emitter, this advanced wave is absorbed, making the original electron recoil in such a way that it radiates a second advanced wave back into the past. This 'new' advanced wave exactly cancels out the 'original' advanced wave, so that there is no effective radiation going back into the past before the moment when the original emission occurred. All that is left is a double wave linking the emitter and the absorber, made up half of a retarded wave carrying positive energy into the future and half of an advanced wave carrying negative energy into the past (in the direction of negative time). Because two negatives make a positive, this advanced wave adds to the original retarded wave as if it too were a retarded wave travelling from the emitter to the absorber.[1]

In Cramer's words: 'The emitter can be considered to produce an "offer" wave which travels to the absorber. The absorber then returns a "confirmation" wave to the emitter, and the transaction is completed with a "handshake" across spacetime.'[2] But this is only the sequence of events from the point of view of pseudotime. In reality, the process is atemporal; it happens all at once. This is because signals that travel at the speed of light take no time at all to complete any journey – in effect, for light signals every point in the Universe is next door to every other point in the Universe. Whether the signals are travelling

---

[1] The entire argument works just as well if you start with the 'absorber' electron emitting radiation into the past; the transactional interpretation itself says nothing about which direction of time should be preferred, but suggests that this is linked to the boundary conditions of the Universe, which favour an arrow of time pointing away from the Big Bang.

[2] 'Transactional interpretation', p. 661.

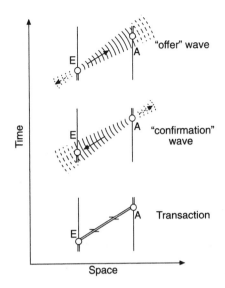

Figure 29

John Cramer's 'transactional interpretation' of quantum mechanics is summed up in this diagram. Reading from the top down, an emitter E sends out an 'offer wave' into the future and into the past (top). This is picked up by an absorber A, which sends an echoing 'confirmation wave' backwards in time to the emitter and into the future (middle). The offer wave and the confirmation wave cancel each other out everywhere in the Universe except in the direct path between the absorber and emitter, where they reinforce one another to produce a quantum transaction (bottom). This diagram is all you need to understand in order to explain all of the quantum mysteries. It is the myth for our times.

backwards or forwards in time doesn't matter, since they take zero time (in their own frame of reference), and $+0$ is the same as -0.

The situation is more complicated in three dimensions, but the conclusions are exactly the same. Taking the most extreme possible case, in a universe which contained just a single electron, the electron would not be able to radiate at all (nor, if Mach's Principle is correct, would it have any mass). If there were just one other electron in the universe, the first electron would be able to radiate, but only in the direction of this second 'absorber' electron. In the real Universe, if matter were not distributed uniformly on the largest scales, and there were less potential for absorption in some directions than in others, we would find that emitters (such as radio antennas) would 'refuse' to radiate equally strongly in all directions. Attempts have actually been made to test this possibility by beaming microwaves out into the Universe in different directions, but they show no sign of any reluctance of the electrons to radiate in any particular direction.

Cramer is at pains to stress that his interpretation makes no predictions that are different from those of conventional quantum mechanics, and that it is offered as a conceptual model which may help people to think clearly about what is going on in the quantum world, a tool which is likely to be particularly useful in teaching, and which has considerable value in developing intuitions and insights into otherwise mysterious quantum phenomena. But there is no need to feel that the transactional interpretation suffers in comparison with other interpretations in this regard, because, as we have seen, none of them is anything other than a conceptual model designed to help our understanding of quantum phenomena, and all of them make the same predictions. The *only* valid criterion for choosing one interpretation rather than another is how effective it is as an aid to our way of thinking about these mysteries – and on that score Cramer's interpretation wins hands down.

First, it not only offers something rather more than a hint of why there is an arrow of time, it also puts all physical processes on an equal footing. There is no need to assign a special status to the observer (intelligent or otherwise), or to the measuring apparatus. At a stroke, this removes the basis for a large part of the philosophical debate about the meaning of quantum mechanics that has gone on for more than half a century. And, going beyond the debate about the role of the observer, the transactional interpretation really does resolve those classic quantum mysteries. I'll give just a couple of examples – how Cramer deals with the experiment with two holes, and how his interpretation makes sense of the Aspect experiment.

If we are going to explain the central mystery of the experiment with two holes, we might as well go the whole hog and explain the ultimate version of this mystery, John Wheeler's variation on the theme, the 'delayed–choice' experiment discussed in Chapter Three. In one version of this experiment, remember, a source of light emits a series of single photons which travel through the experiment with two holes. On the other side is a detector screen which can record the positions the photons arrive at, but which can be flipped down, while the photons are on their way, to allow them to pass on to one or other of a pair of telescopes focused on the two slits (one focused on each slit). If the screen is down, the telescopes will observe single photons each passing through one or other of the slits, with no sign of interference; if the screen is up, the photons will seem to pass through both slits, creating an interference pattern on the screen. And the screen can be flipped down *after* the photons have passed the slits, so that their decision about which pattern of behaviour to adopt seems to be determined by an event which occurs after they have made that decision.

In Cramer's version of events, a retarded 'offer wave' (monitored in 'pseudo-time' for the purpose of this discussion) sets off through both holes in the experiment. If the screen is up, the wave is absorbed in the detector, triggering an advanced 'confirmation wave' which travels back through *both* slits of the apparatus to the source. The final transaction forms along both possible paths (actually, as Feynman would have stressed, along *every* possible path), and there is interference.

If the screen is down, the offer wave passes on to the two telescopes trained on the slits. Because each telescope is trained on just one slit, it is only possible for any confirmation wave produced when the offer wave interacts with the telescope itself to go back to the source through the slit on which that telescope is trained. And, of course, the absorption event must involve a whole photon, not a part of a photon. Although each telescope may send back a confirmation wave through its respective slit, the source has to 'choose' (at random) which one to accept, and the result is a final transaction which involves the passage of a single photon through a single slit. The evolving state vector of the photon 'knows' whether the screen is going to be up or down because the confirmation wave really does travel back in time through the apparatus, but the whole transaction is, as before, atemporal.

> The issue of when the observer decides which experiment to perform is no longer significant. The observer determined the experimental configuration and boundary conditions, and the transaction formed accordingly. Furthermore, the fact that the detection event involves a measurement (as opposed to any other interaction) is no longer significant, and so the observer has no special role in the process.[1]

You can amuse yourself by working out a similar explanation of what happens to Schrödinger's cat (and Wigner's friend). Once again, what matters is that the completed transaction only allows one possibility (dead cat or live cat) to become real, and because the 'collapse of the wave function' does not have to wait for the observer to look into the box, there is never a time when the cat is half dead and half alive. It's a sign of how powerful and straightforward the transactional interpretation is that I am sure you can indeed work out the details for yourself, without me spelling them out.

But what about Bell's inequality, the Einstein–Podolsky–Rosen

---

[1] Cramer, p. 673.

paradox, and the Aspect experiment? This, after all, was what revived interest in the meaning of quantum mechanics in the 1980s. From the point of view of absorber theory, there is no difficulty in understanding what is going on. We imagine (still thinking in terms of pseudo-time) that the excited atom which is about to emit two photons sends out offer waves in various directions and corresponding to various possible polarization states. The transaction is completed, and the photons actually emitted, only if confirmatory advanced waves are sent back in time from the appropriate pair of observers to the emitting atom. As soon as the transaction is complete, the photons are emitted and observed, producing a double detection event in which the polarizations of the photons are correlated, even though they are far apart in space. If the confirmatory waves do not match an allowed polarization correlation, then they cannot be 'verifying' the same transaction, and they will not be able to establish the handshake. From the perspective of pseudo-time, the pair of photons *cannot* be emitted until an arrangement has been made to absorb them, and that absorption arrangement itself determines the polarizations of the emitted photons, even though they are emitted 'before' the absorption takes place. It is literally impossible for the atom to emit photons in a state that does not match the kind of absorption allowed by the detectors. Indeed, in the absorber model the atom cannot emit photons at all unless an agreement has already been reached to absorb them.

It's the same with those two kittens travelling in their separate spacecraft to the opposite ends of the Galaxy. The observation that determines which half-box the electron is in, and therefore which kitten lives and which kitten dies, echoes backwards in time to the start of the experiment, instantaneously (or rather, atemporally) determining the states of the kittens throughout the entire period when they were locked away, unobserved, in their spaceships.

> If there is one particular link in [the] event chain that is special, it is not the one that ends the chain. It is the link at the beginning of the chain when the emitter, having received various confirmation waves from its offer wave, reinforces one of them in such a way that it brings that particular confirmation wave into reality as a completed transaction. The atemporal transaction does not have a 'when' at the end.[1]

This dramatic success in resolving all the puzzles of quantum physics

[1] Cramer, 1986, p. 674.

has been achieved at the cost of accepting just one idea that seems to run counter to common sense – the idea that part of the quantum wave really can travel backwards through time. At first sight, this is in stark disagreement with the common sense intuition that causes must always precede the events that they cause. But on closer inspection it turns out that the kind of time travel required by the transactional interpretation does not violate the everyday notion of causality after all – and nor does all of this atemporal handshaking across the Universe necessarily remove that most prized of our human attributes, our freedom of will.

TAKING TIME TO MAKE TIME

In the everyday world, it is obvious that effects always follow their causes. I work out what the next sentence is going to be in my head, then I tap away at the keys on my computer, and a tiny fraction of a second after I hit each key the relevant letter appears on the screen. It is not (alas) true that the words appear on the screen first, and that I then read them to work out what it is that I want to say. But when an atemporal handshake takes place with the aid of an advanced quantum wave that travels backwards in time, this need not have any influence on the logical pattern of causality in the everyday world.

Cramer suggests that there are two kinds of causality, which he dubs 'strong' and 'weak'. The 'weak-causality principle' applies in the everyday world (the 'macroscopic' world), and is the basis of our common-sense ideas about time. It says that a macroscopic cause must always precede its macroscopic effects in any reference frame. Macroscopic information can never be transmitted faster than light, or backwards in time. Most people would go along with that. But Cramer also defines a 'strong-causality principle' which says that a cause must always precede *all* of its effects in any reference frame, so that even on the microscopic scale (that is, the quantum scale) information cannot be transmitted backwards in time or faster than light. This is often taken to be an obvious extension of the weak-causality principle; but Cramer points out that there is actually no experimental evidence for strong causality. Indeed, such experimental evidence as there is – the tests of Bell's inequality – explicitly shows that 'microscopic' causality is violated, regardless of which interpretation of quantum mechanics you favour. In absorber theory, there are always violations of strong causality; but there is no violation of weak causality as long as absorption is always complete in the direction of the future.

It should be no surprise that the way the transactional interpretation deals with time differs from common sense, because the transactional interpretation explicitly includes the effects of relativity theory, and we have already seen how non-commonsensical they are when it comes to describing time. The Copenhagen Interpretation, by contrast, treats time in the classical, 'Newtonian' way, and this is at the heart of the inconsistencies in any attempt to explain the results of experiments like the Aspect experiment in terms of the Copenhagen Interpretation. If the velocity of light were infinite, the problems would disappear; there would be no difference between the local and non-local descriptions of processes involving Bell's inequality, and the ordinary Schrödinger equation would be an accurate description of what is going on – the ordinary Schrödinger equation is, in effect, the correct 'relativistic' equation when the speed of light is infinite. Cramer has actually found a rather subtle link between relativity and quantum mechanics, and this is at the heart of his interpretation.

How does the atemporal handshaking affect the possibility of free will? At first sight, it might seem as if everything is fixed by these communications between the past and the future. Every photon that is emitted already 'knows' when and where it is going to be absorbed; every quantum probability wave, slipping at the speed of light through the slits in the experiment with two holes, already 'knows' what kind of detector is waiting for it on the other side. We are back with the image of a frozen Universe, viewed from the perspective of a photon, in which neither time nor space has any meaning, and everything that ever was or ever will be just *is*.

But this, remember, is the perspective of a photon, or of anything else (such as a quantum probability wave) that travels at the speed of light. For macroscopic objects like human beings, time is real enough. In my frame of reference, I still have time to decide what the next sentence is going to be, and whether to take a break for lunch now or in twenty minutes' time. The decisions I make may produce an interlocking web of atemporal quantum connections, so that a photon, if it could speak, would be able to tell me how those decisions are going to affect my future life; but the weak-causality principle protects me from any such leakage of information from the microscopic world to the macroscopic world. In my time-frame those decisions are made with genuine free will and no certain knowledge of their outcomes. It takes time (in the macroscopic world) to make the decisions (both human decisions and quantum 'choices' like those involved in the decay of an atom) which make the atemporal reality of the microscopic world. What we experience is much more like Cramer's 'pseudo-time' than like the atemporal handshaking that underlies quantum interactions.

At least, that's how I see it. Like everything else in the story, this is just an analogy, a myth or a model. You may find another way of thinking about the way our everyday time sense interlocks with the atemporal quantum world. You might even prefer to accept, following John Bell's mischievous suggestion, that there is no such thing as free will after all, and that the success of the transactional interpretation is evidence that everything is preordained (from our human perspective), and that I had no choice about writing this book and you had no option but to read it. But although the non-locality of the Universe at the microscopic level may make us uncomfortable, and may make it hard to understand the relationship between past, present and future in everyday terms, remember that this is not a unique feature of the transactional interpretation. It is an experimental fact, which has to be taken account of in any satisfactory interpretation of quantum reality. Furthermore, this atemporal linking of different parts of spacetime into a coherent whole seems to match rather well the image from relativity theory of a continuous spacetime 'history', discussed in Chapter Two. The success of the transactional interpretation rests largely upon the way in which it confronts this issue squarely, and builds outward from the atemporality of the quantum world revealed by the experiments to test Bell's inequality.

I stress, again, that *all* such interpretations are myths, crutches to help us imagine what is going on at the quantum level and to make testable predictions. They are not, any of them, uniquely 'the truth'; rather, they are *all* 'real', even where they disagree with one another. But Cramer's interpretation is very much a myth for our times; it is easy to work with and to use in constructing mental images of what is going on, and with any luck at all it will supersede the Copenhagen Interpretation as the standard way of thinking about quantum physics for the next generation of scientists.

It is certainly a superb way of teaching quantum physics to beginners (that is, to anyone who hasn't already been corrupted by the Copenhagen Interpretation). As Cramer says:

> A shift away from the Copenhagen Interpretation may be particularly difficult because of its traditional role in the teaching of quantum mechanics over five decades.
>
> The value of new interpretational insights into physical processes, however, should not be underestimated. Experience in many fields of physics has shown that progress and new ideas and approaches are stimulated by the ability to visualize clearly physical phenomena.[1]

Back in 1977, discussing the difficulty of understanding the outcome of quantum experiments in terms of interactions that, in principle, involve the whole Universe, Fred Hoyle commented: 'Success may come one day, however, but only from a nonlocal form of physics, the kind of physics that is not at all popular right now.'[2] Hoyle's prescient comment, and Cramer's hope, show signs of being fulfilled in work like that of Chu on the nature of gravity. This is not the end of the story of quantum mechanics, but the beginning of a new chapter in the story of quantum mechanics. But there is one last irony with which I want to close this account of the story so far.

Of all the great physicists of the twentieth century, the one who most clearly and frequently expressed the essential incomprehensibility of quantum mechanics in its standard form was Richard Feynman. In the mid-1960s, for example, he wrote in his book *The Character of Physical Law*:

> There was a time when the newspapers said that only twelve men understood the theory of relativity. I do not believe that there ever was such a time. There might have been a time when only one man did, because he was the only guy who caught on, before he wrote his paper. But after people read the paper a lot of people understood the theory of relativity in some way or other, certainly more than twelve. On the other hand, I think I can safely say that nobody understands quantum mechanics. ... Do not keep saying to yourself, if you can possibly avoid it, 'But how can it be like that?' because you will go 'down the drain' into a blind alley from which nobody has yet escaped. Nobody knows how it can be like that.[3]

---

[1] 1986, p. 681.

[2] Hoyle, *Ten Faces of the Universe* (London: Heinemann, 1977), p. 128.

[3] P. 129; London: BBC Publications, 1965 (based on lectures given in 1964; reprinted by MIT Press in 1967 and many times since).

The irony, of course, is that the means of escape from that blind alley builds from the theory of light that Feynman himself had come up with 20 years before he made that comment. But it has taken a further 30 years for this to become clear. It may only be a myth for our times, but the great thing about John Cramer's transactional interpretation is that it does allow you to ask the question 'But how can it be like that?' and to come up with a simple and easily understood answer which does not involve going 'down the drain'. What more can you ask of any interpretation of quantum mechanics?

BIBLIOGRAPHY

As well as the specific references in the text, which are usually to more technical books and scientific papers, these are the books that I found particularly useful (in some cases, influential) in developing my ideas about the meaning of quantum reality and what physics is all about. I have included several of my own books in this bibliography, because they show how my own ideas have developed and changed over the past two decades.

David Albert, *Quantum Mechanics and Experience* (Cambridge, Mass.: Harvard University Press, 1992).
Argues the case for the 'many-minds' interpretation of quantum mechanics, but leaves me completely unconvinced. If you are intrigued by the idea, this is the place to find out how convincing the argument seems to you.

Hans von Baeyer, *Taming the Atom* (London: Viking, 1992).
Gives you a good feel for the world of atoms and molecules, and includes stunning photographs of single atoms, DNA molecules and other wonders of the microworld. But beware of some mistakes, including the 'explanation' of the structure of the helium atom.

Jim Baggott, *The Meaning of Quantum Theory* (Oxford: Oxford University Press, 1992).
A somewhat technical account written by a physicist who had been startled to discover Bell's theorem as late as 1987, having previously lived his life in blissful ignorance of the significance of quantum non-locality; his naive sense of wonder at all the mysteries he has just discovered gives the book great appeal, if you read around the equations.

Ralph Baierlein, *Newton to Einstein* (Cambridge: Cambridge University Press, 1992).
Pitched at undergraduates who are not science specialists (and therefore relatively intelligible to anyone interested in the subject), this book deals with the dual nature of light as particle and wave, and outlines the special theory of relativity. It is still a textbook, but much more accessible than most textbooks.

J. S. Bell, *Speakable and Unspeakable in Quantum Mechanics* (Cambridge: Cambridge University Press, 1987).
A collection of *all* of John Bell's papers on the conceptual and philosophical problems of quantum theory. Some quite accessible, some highly technical.

Paul Davies, *Other Worlds* (London: Pelican, 1988; original edition London: J. M. Dent, 1980).
A good, slightly dated, overview of quantum ideas, written before the Aspect experiment was carried out. Davies presents a favourable overview of the 'many-worlds' theory, and discusses the anthropic 'coincidences' that make the world the way it is.

Paul Davies and J. R. Brown (eds), *The Ghost in the Atom* (Cambridge: Cambridge University Press, 1986).
'Horse's mouth' versions of different interpretations of the meaning of quantum theory, based on interviews for a BBC Radio series. Eminent experts argue for mutually incompatible possibilities, all on the basis of the same evidence! A fine example of the confusion surrounding physicists' understanding of the meaning of quantum mechanics.

David Deutsch, *The Fabric of Reality* (London: Viking, 1995).
A very personal view of quantum reality developed from Hugh Everett's 'many-worlds' theory and including some intriguing ideas about the nature of time.

J. W. Dunne, *An Experiment with Time*, 3rd edn (London: Faber & Faber, 1934).
A slightly mystic discussion of the nature of time which brings out clearly the need for a second layer of time in order to measure the 'flow' of everyday time, a third layer of time to measure the second layer, and so on *ad infinitum*.

C. W. F. Everitt, *James Clerk Maxwell* (New York: Scribner's, 1975).
A readable and straightforward account of Maxwell's life and work.

J. Fauvel, R. Flood, M. Shortland and R. Wilson (eds), *Let Newton Be!* (Oxford: Oxford University Press, 1988).
A very accessible collection of articles about Newton and his work.

Richard Feynman, *QED: The strange theory of light and matter* (London: Penguin, 1990).
The latest printing I have of a book first published in 1985 and based on a series of lectures given by Feynman in 1983 to a non-scientific audience in Los Angeles. A wonderful example of Feynman's pictorial way of explaining how quantum physics 'works'.

Richard Feynman, *The Character of Physical Law* (London: Penguin, 1992).
A new edition of a book first published in 1965 and based on a series of lectures broadcast by the BBC. Includes a chapter on quantum theory, but the whole book is well worth reading – the authentic Feynman 'voice'.

Richard Feynman, *Six Easy Pieces* (Mass.: Addison-Wesley, 1995).
Six of the introductory lectures from Feynman's famous physics course (see below), including an introduction to quantum physics.

Richard Feynman, Robert Leighton and Matthew Sands, *The Feynman Lectures on Physics, Vol. III* (Mass.: Addison-Wesley, 1965).
The volume of Feynman's famous lectures that deals with quantum theory. An undergraduate text, for anyone seriously interested in the subject.

Richard Feynman and Steven Weinberg, *Elementary Particles and the Laws of Physics* (Cambridge: Cambridge University Press, 1987).
Transcripts of two lectures given in Cambridge in the mid-1980s in honour of Paul Dirac. Very good at providing a feel for how physicists think.

Kathleen Freeman, *Ancilla to the Pre-Socratic Philosophers* (Cambridge, Mass.: Harvard University Press, 1983).
Includes the portions of the work of Empedocles referred to in Chapter One.

James Gleick, *Genius* (London: Little Brown, 1992).
A comprehensive study of Feynman's life and work, set in the context of twentieth-century physics.

John Gribbin, *In Search of Schrödinger's Cat* (New York: Bantam, and London: Black Swan, 1984).
Leaves off where the book in front of you begins. The best layperson's guide to the story of how quantum theory emerged (but I would say that, wouldn't I?).

John Gribbin, *In Search of the Big Bang* (New York: Bantam, and London: Black Swan, 1986).
The standard theory of the origin of the Universe, set in the context of ideas about quantum physics.

John Gribbin, *In Search of the Edge of Time* (New York: Harmony, and London: Black Swan, 1992).
My account of the emergence of relativity theory, and its implications, including the understanding of time and the possibility of time travel.

John Gribbin, *In the Beginning* (New York: Little, Brown, and London: Viking, 1993).
Latest ideas about the origin of the Universe, and the evidence that it is 'closed' in the way that matches the requirements of Wheeler–Feynman absorber theory.

John and Mary Gribbin, *Time and Space* (London: Dorling Kindersley, 1994).
An attempt to provide a simple explanation of Einstein's theories of relativity in an accessible, heavily illustrated format with minimum text. May help to clarify some of the ideas expressed in Chapter Two!

Herman Haken, Anders Karlqvist and Uno Svedin (eds), *The Machine as Metaphor and Tool* (Berlin: Springer-Verlag, 1993).
A collection of articles, developed from a workshop held in Abisko, Sweden, in May 1990, around the theme of the machine and its use as a metaphor in a variety of contexts, including the scientific world view. Mainly about the brain, but relevant to the themes I discuss in Chapter Five.

Nick Herbert, *Quantum Reality* (London: Rider, 1985).
A very readable, slightly out-of-date exposition of different interpretations of quantum theory.

Roger Jones, *Physics as Metaphor* (Minneapolis, Minn.: University of Minnesota Press, 1982).
Looks at the way physicists think about the world, and questions commonplace assumptions about the relationships between models and reality.

Martin Krieger, *Doing Physics* (Bloomington, Ind.: Indiana University Press, 1992).
A mind-expanding book which brings home more clearly and forcefully than any other discussion I know of the extent to which physics is not just based on but *is* a system of analogies and metaphors – in other words, fiction. Closely reasoned, and needs careful study; but if you make the effort you will never see the world of science in the same light again.

Thomas Kuhn, *The Structure of Scientific Revolutions* (Chicago: University of Chicago Press, 1970).
A classic work about the way scientists work and think – and how and why they sometimes change their minds.

Jean-Pierre Maury, *Newton: Understanding the Cosmos* (London: Thames & Hudson, 1992).
English translation of a French book that first appeared in 1990. By far the best 'instant guide' to Newton and his work, with easy-to-read text, colourful illustrations, and all in 144 pocket-sized pages.

Dugald Murdoch, *Niels Bohr's Philosophy of Physics* (Cambridge: Cambridge University Press, 1987).
A scholarly assessment of Bohr's contribution to quantum theory, and especially of what exactly he meant by what is now known as the Copenhagen Interpretation. Not always an easy read, but the place to go for the real nitty-gritty.

Heinz Pagels, *The Cosmic Code* (London: Michael Joseph, 1982).
A clear and interesting account of the strangeness of the quantum world (and especially the Copenhagen Interpretation), written just before the wave of interest in alternative interpretations inspired by the results of the Aspect experiment, by an eminent physicist with a gift for communication.

Roger Penrose, *The Emperor's New Mind* (Oxford: Oxford University Press, 1989).
In order to build up a case that truly intelligent computers cannot exist, Penrose takes the reader on a tour through most of modern physics, including quantum theory. Heavy going in parts, an exhilarating read in others, often contentious, but well worth reading.

Andrew Pickering, *Constructing Quarks* (Edinburgh: Edinburgh University Press, 1984).
A fascinating account of the history of modern particle physics, heavy going in places, but presenting the story, and the final (?) theory, as a result not of scientists uncovering a hidden truth, but creating the reality out of their experiments and theories. Amply repays careful reading.

William Poundstone, *Labyrinths of Reason* (New York: Anchor Books, 1988).
An accessible look at the way physicists think about the world.

Ilya Prigogine and Isabelle Stengers, *Order out of Chaos* (London: Heinemann, 1984).
A good introduction to Prigogine's ideas about complexity and the arrow of time, but very heavy going in places. An even heavier version is to be found in Prigogine's solo effort *From Being to Becoming* (San Francisco: Freeman, 1980). A new book by Prigogine and Stengers is due to be published in 1995.
   Prigogine's ideas have been elaborated in a series of books which are packed with intriguing ideas, but which I find rather heavy going in places. Fortunately, the possible relevance of these ideas to the quantum world has been discussed particularly clearly by Alastair Rae, in his book *Quantum Physics: Illusion or Reality?* (see below), which I recommend for a quick overview.

Alastair Rae, *Quantum Physics: Illusion or Reality?* (Cambridge: Cambridge University Press, 1986).
A standard, fairly traditional, layman's guide; includes a discussion of Ilya Prigogine's work which is more accessible than Prigogine's own books.

Henry Stapp, *Mind, Matter, and Quantum Mechanics* (Berlin: Springer-Verlag, 1993).
Tough going in places, this books benefits from being a collection of articles by Stapp, all addressing the questions of quantum theory and consciousness. Because his ideas are presented several times in slightly different ways, the persistent reader can eventually get a taste of what is going on. Worth the effort if you want to probe deeper into the mystery of mind and matter which I touched on in Chapter Four.

John Tyndall, *On Light* (London: Longman, 1873).
A delightful book, based on lectures given by Tyndall on a tour of the United States. An intriguing window on the world of Victorian science, opened for

us by the man who first realized why the sky is blue. He explains his ideas on p. 152 of this volume.

Robert Weber, *Pioneers of Science*, 2nd edn (Bristol: Adam Hilger, 1988).
Thumbnail sketches of every Nobel prizewinner in physics from the first (Wilhelm Röntgen, in 1901) to Alex Müller and Georg Bednorz in 1987.

Richard Westfall, *Never at Rest* (Cambridge: Cambridge University Press, 1980).
The definitive biography of Newton. A cut-down version of the same book, retitled *The Life of Isaac Newton*, was published by CUP in 1993, and might be easier to get hold of; but the original is far better.

John Wheeler and Wojciech Zurek, *Quantum Theory and Measurement* (Princeton: Princeton University Press, 1983).
A superb collection of reprints of the classic papers in the history of the investigation of the meaning of quantum theory. The EPR paper, the first appearance of Schrödinger's cat, Bohm, Bell and Aspect are all here, along with many others (but stopping short, alas, of Cramer) and a small amount of commentary. Mostly highly technical; worth dipping into in a library.

Arthur Zajonc, *Catching the Light* (London: Bantam, 1993).
A fascinating look at the history of light, including artistic and poetic impressions as well as science.

INDEX